Advances in Integrated Pest
and
Disease Management in
Horticultural Crops

Advances in Integrated Pest and Disease Management in Horticultural Crops

Vol. 3

Dr. P. Singh

AGROTECH PRESS

JAIPUR ▪ NEW DELHI

AGROTECH PRESS

Regd. Off. "Bony Residency", Gate No. 2
Opp. Tilak Public School, Vishweswariya Nagar,
Gopalpura Road, Jaipur - 302018, India
Ph: 0141 - 2761280, Telefax: 2761381
E-mail : agrotechpress@gmail.com

Delhi Office: 102, 1st Floor, Satyam House,
4327/3, Ansari Road, Darya Ganj,
New Delhi - 110002, India
Ph: 011-45652440
visit us: www.agrotechpress.com

Distributed By
KEYA INTERNATIONAL PRESS LLP
E-mail: keyainternationalpressllp@gmail.com

Published in India, 2024

ISBN : 9789384568535 (4 Vols. Set)

Typeset by Thomson Press

Printed and bound in India by Thomson Press Ltd.,
New Delhi

Contents

Preface

Integrated Pest Management (IPM) is an approach to managing insects, mites, pathogens, nematodes, weeds, and other pests in which multiple practices are implemented throughout the entire production period of the crop. IPM can be viewed as a series of steps that is repeated or modified as needed: Prevention/avoidance, Monitoring, Decision making, Intervention, Evaluation. Floriculture and nursery managers are increasingly adopting IPM principles and tactics. Benefits of IPM methods include reducing pesticide resistance; minimizing phytotoxicity and disruptions that occur from pesticide reentry intervals; and reducing the costs of pesticide purchases, application labour, and regulatory compliance. Of the several factors responsible for lower productivity of ornamental, medicinal, aromatic and tuber crops, pests are considered as important limiting factors. The book deals with integrated pest, disease and nematode management in ornamental, medicinal, aromatic and tuber crops in detail using regulatory, physical, cultural, chemical, biological, host plant resistance and integrated methods.

A key risk horticultural business related to ornamental crops face is the potential damage that can be caused by pests and diseases during production and/or storage. The tolerance for damage in the ornamentals sector is particularly low, given the fact that minuscule levels of cosmetic damage can result in crops being rejected by end users. Economic, environmental, and technological influences complicate the task of achieving disease-free products in the ornamentals industry. Integrated pest management (IPM) is a cornerstone of floriculture and nursery crop production: strategies include sanitation, clean stock, host resistance, and control through biological, cultural, environmental, chemical, and regulatory means. Sanitation measures and cultural controls must keep pace with new production technologies. Clean stock programs are used for many crops that are propagated vegetatively. Breeding, selection, and biotechnology provide crops resistant to pathogens. Offshore production for economic competitiveness can introduce pathogens that make regulatory programs necessary. New biocontrol and chemical products continue to improve control while meeting the requirement for minimal environmental impact. Continual introduction of new crops and new production technologies creates new opportunities for pathogens to exploit, such that new disease management tactics must be discovered and old ones rediscovered to achieve optimum health management for ornamentals. The book provides information on pests affecting bulbs, cut flowers, potted flowering plants, foliage plants, bedding plants, and ornamental trees and shrubs grown in the field, greenhouse, and nursery. It discusses a range of options for the control of priority pests and diseases on ornamental plants and for optimising nursery hygiene.

Cultivation of medicinal plants should also ensure maintaining standards - in quality, potency and chemical composition of the produce. The active principles of the plant are generally secondary metabolites and their biosynthesis, though controlled genetically, is strongly affected by soil, climatic and agricultural factors. It is, therefore advised not to use chemical fertilizers, insecticides and pesticides in cultivation of medicinal plants. Application of chemical fertilizers may alter the desired chemical composition of medicinal plants and application of herbicides and other plant protection chemicals may lead to accumulation of their residue in the plant. It will affect the chemical composition of the plant, which may cause great health hazards to the human being, besides polluting the environment. The employment of Integrated Pest and Disease Management in medicinal and aromatic plants shall ensure sustainable development not only in terms of production of medicinal plants, but also in socio-economic and ecological parameters. Root and tuber vegetables include potatoes, carrots, parsnips, turnips, beetroot, sweetpotato, taro etc. They are members of a number of different vegetable families. A range of insect pests, disease and nematodes can cause economic losses in root and tuber vegetables. Additionally, invertebrate pests such as slugs and snails can be a problem. The book provides an overview of the key pests and diseases of root and tuber vegetables. The related strategies of IPM for tuber crops can be used by growers and crop consultants to accurately identify the range of pests and diseases encountered in root and tuber vegetables. Such strategies will result in better safety for applicators and the food supply and will minimize environmental contamination.

The book covers various aspects of disease and pest management as they relate specifically to ornamental, medicinal, aromatic and tuber plants. IPM aims to ensure the most efficient use of control methods to: minimise crop damage and losses; costs of control measures; and negative environmental impacts. IPM makes good business sense. As discussed earlier IPM is not the routine use of chemical control measures that is seen in many commercial crop production protocols. The subject matter in this book should attract the reader's concern to support rational decisions regarding various Integrated Pest and disease Management strategies in horticulture.

Dr. P. Singh

1

Ornamental Crops

Ornamental plants are plants that are grown for decorative purposes in gardens and landscape design projects, as houseplants, for cut flowers and specimen display. The cultivation of these, called floriculture, forms a major branch of horticulture.

Garden Plants

Most commonly ornamental garden plants are grown for the display of aesthetic features including: flowers, leaves, scent, overall foliage texture, fruit, stem and bark, and aesthetic form. In some cases, unusual features may be considered to be of interest, such as the prominent and rather vicious thorns of *Rosa sericea* and *cacti*. In all cases, their purpose is for the enjoyment of gardeners, visitors, and/or the public.

Trees

Similarly certain trees may be called ornamental trees. This term is used when they are used as part of a garden or landscape setting, for instance for their flowers, their texture, form and shape, and other aesthetic characteristics. In some countries trees in 'utilitarian' landscape use such as screening, and roadside plantings are called *amenity trees*.

Cultivation

For plants to be considered to be ornamental, they may require specific work and activity by a gardener. For instance, many plants cultivated for topiary and bonsai would only be considered to be ornamental by virtue of the regular pruning carried out on them by the gardener, and they may rapidly cease to be ornamental if the work was abandoned.

Ornamental plants and trees are distinguished from utilitarian and crop plants, such as those used for agriculture and vegetable crops, and for forestry or as fruit trees. This does not preclude any particular type of plant being grown both for ornamental qualities in the garden, and for utilitarian purposes in other settings. Thus lavender is typically grown as an ornamental plant in gardens, but may also be grown as a crop plant for the production of lavender oil.

The term ornamental plant is used here in the same sense that it is generally used in the horticultural trades. The term largely corresponds to 'garden plant', though the latter is much less precise, as any plant may be grown in a garden. Ornamental plants are plants which are grown for display purposes, rather than functional ones. While some plants are both ornamental and functional, people usually use the term "ornamental plants" to refer to plants which have no value beyond being attractive, although many people feel that this is value enough. Ornamental plants are the keystone of ornamental gardening, and they come in a range of shapes, sizes and colours suitable to a broad array of climates, landscapes, and gardening needs.

Some ornamental plants are grown for showy foliage. Their foliage may be deciduous, turning bright orange, red, and yellow before dropping off in the fall, or evergreen, in which case it stays green year round. Some ornamental foliage has a striking appearance created by lacy leaves or long needles, while other ornamentals are grown for distinctively coloured leaves, such as silvery-gray groundcovers and bright red grasses, among many others.

Other ornamental plants are cultivated for their blooms. Flowering ornamentals are a key aspect of many gardens, with many flower gardeners preferring to plant a variety of flowers so that the garden is continuously in flower through the spring and summer. Depending on the types of plants being grown, the flowers may be subtle and delicate, or large and showy, with some ornamental plants producing distinctive aromas which paint a palette of scents in addition to colours.

Ornamental Grass

Ornamental grasses are grasses grown as ornamental plants. They have become increasingly popular in gardens in recent years.

Along with true grasses (Poaceae), several other families of grass-like plants are typically marketed as ornamental grasses. These include the sedges (Cyperaceae), rushes (Juncaceae), restios (Restionaceae), and cat-tails (Typhaceae). All are monocotyledons, typically with narrow leaves and parallel veins. Most are herbaceous perennials, though many are evergreen and some develop woody tissues. They bring striking linear form, texture, colour, motion, and sound to the garden, throughout the year.

Almost all ornamental grasses are perennials, coming up in spring, from their roots, which have stored large quantities of energy, and in fall or winter go dormant. Some, notably bamboos, are evergreen, and a few are annuals. Many are bunch grasses and tussock grasses, though others form extensive systems of many-branched rhizomes. The bunching types are often called "clump-forming" or "clumping", distinct from the rhizomatous types, called "running". Sizes vary from a few centimetres up to several metres; the larger bamboos may reach 20 m or more tall. Some ornamental grasses are species that can be grown from seed. Many others are cultivars, and must be propagated by vegetative propagation of an existing plant.

Pampas grass (*Cortaderia selloana*) is easily recognizable, with semi-dwarf to very large selections for the landscape. Deer grass (*Muhlenbergia rigens*) and canyon prince wild blue rye (*Leymus condensatus*) are popular in larger settings, natural landscaping, and native plant gardens. There are *Miscanthus* grasses whose variegations are horizontal, and appear even on a cloudy day to be stippled with sunshine. Many *Miscanthus* and *Pennisetum* species flower in mid or late summer, and the seed heads are long lasting, often remaining well into the winter. Some *Stipa* species flower in the spring, the inflorescence standing almost two metres above the clumps of leaves, and again the seed heads last late into the winter.

When gardening near natural wildland-urban interfaces, care to avoid invasive species, such as *Cortaderia jubata, Pennisetum setaceum*, and *Stipa tenuissima* (syn. *Nassella tenuissima*), is responsible horticulture.

Ornamental Flowers

A flower, sometimes known as a bloom or blossom, is the reproductive structure found in flowering plants (plants of the division Magnoliophyta, also called angiosperms). The biological function of a flower is to effect reproduction, usually by providing a mechanism for the union of sperm with eggs. Flowers may facilitate outcrossing (fusion of sperm and eggs from different individuals in a population) or allow selfing (fusion of sperm and egg from the same flower). Some flowers produce diaspores without fertilization (parthenocarpy). Flowers contain sporangia and are the site where gametophytes develop. Flowers give rise to fruit and seeds. Many flowers have evolved to be attractive to animals, so as to cause them to be vectors for the transfer of pollen.

In addition to facilitating the reproduction of flowering plants, flowers have long been admired and used by humans to beautify their environment, and also as objects of romance, ritual, religion, medicine and as a source of food.

Symbolism

Figure: *Red Rose*

Figure: *Lilies are often used to denote life or resurrection*

Many flowers have important symbolic meanings in Western culture. The practice of assigning meanings to flowers is known as floriography. Some of the more common examples include:

- Red roses are given as a symbol of love, beauty, and passion.
- Poppies are a symbol of consolation in time of death. In the United Kingdom, New Zealand, Australia and Canada, red poppies are worn to commemorate soldiers who have died in times of war.
- Irises/Lily are used in burials as a symbol referring to "resurrection/life". It is also associated with stars (sun) and its petals blooming/shining.
- Daisies are a symbol of innocence.

Flowers within art are also representative of the female genitalia, as seen in the works of artists such as Georgia O'Keeffe, Imogen Cunningham, Veronica Ruiz de Velasco, and Judy Chicago, and in fact in Asian and western classical art. Many cultures around the world have a marked tendency to associate flowers with femininity.

The great variety of delicate and beautiful flowers has inspired the works of numerous poets, especially from the 18th-19th century Romantic era. Famous examples include William Wordsworth's *I Wandered Lonely as a Cloud* and William Blake's *Ah! Sun-Flower*.

Because of their varied and colourful appearance, flowers have long been a favourite subject of visual artists as well. Some of the most celebrated paintings from well-known painters are of flowers, such as Van Gogh's sunflowers series or Monet's water lilies. Flowers are also dried, freeze dried and pressed in order to create permanent, three-dimensional pieces of flower art.

Their symbolism in dreams has also been discussed, with possible interpretations including "blossoming potential".

The Roman goddess of flowers, gardens, and the season of Spring is Flora. The Greek goddess of spring, flowers and nature is Chloris.

In Hindu mythology, flowers have a significant status. Vishnu, one of the three major gods in the Hindu system, is often depicted standing straight on a lotus flower. Apart from the association with Vishnu, the Hindu tradition also considers the lotus to have spiritual significance. For example, it figures in the Hindu stories of creation.

Usage

In modern times people have sought ways to cultivate, buy, wear, or otherwise be around flowers and blooming plants, partly because of their agreeable appearance and smell. Around the world, people use flowers for a wide range of events and functions that, cumulatively, encompass one's lifetime:

- For new births or christenings
- As a corsage or boutonniere worn at social functions or for holidays
- As tokens of love or esteem
- For wedding flowers for the bridal party, and for decorations for the hall
- As brightening decorations within the home
- As a gift of remembrance for *bon voyage* parties, welcome-home parties, and "thinking of you" gifts
- For funeral flowers and expressions of sympathy for the grieving
- For worshiping goddesses. In Hindu culture adherents commonly bring flowers as a gift to temples

People therefore grow flowers around their homes, dedicate entire parts of their living space to flower gardens, pick wildflowers, or buy flowers from florists who depend on an entire network of commercial growers and shippers to support their trade.

Flowers provide less food than other major plants parts (seeds, fruits, roots, stems and leaves) but they provide several important foods and spices. Flower vegetables include broccoli, cauliflower and artichoke. The most expensive spice, saffron, consists of dried stigmas of a crocus. Other flower spices are cloves and capers. Hops flowers are used to flavour beer. Marigold flowers are fed to chickens to give their egg yolks a golden yellow colour, which consumers find more desirable; dried and ground marigold flowers are also used as a spice and colouring agent in Georgian cuisine. Flowers of the dandelion and elder are often made into wine. Bee pollen, pollen collected from bees, is considered a health food by some people. Honey consists of bee-processed flower nectar and is often named for the type of flower, e.g. orange blossom honey, clover honey and tupelo honey.

Hundreds of fresh flowers are edible but few are widely marketed as food. They are often used to add colour and flavour to salads. Squash flowers are dipped in breadcrumbs and fried. Edible flowers include nasturtium, chrysanthemum, carnation, cattail, honeysuckle, chicory, cornflower, canna, and sunflower. Some edible flowers are sometimes candied such as daisy, rose, and violet (one may also come across a candied pansy).

Flowers can also be made into herbal teas. Dried flowers such as chrysanthemum, rose, jasmine, camomile are infused into tea both for their fragrance and medical properties. Sometimes, they are also mixed with tea leaves for the added fragrance.

Flowers have been used since as far back as 50,000 years in funeral rituals. Many cultures do draw a connection between flowers and life and death, and because of their seasonal return flowers also suggest rebirth, which may explain why many people place flowers upon graves. In ancient times the Greeks would place a crown of flowers on the head of the deceased as well as cover tombs with wreaths and flower petals. Rich and powerful women in ancient Egypt would wear floral headdresses and necklaces upon their death as representations of renewal and a joyful afterlife, and the Mexicans to this day use flowers prominently in their Day of the Dead celebrations in the same way that their Aztec ancestors did.

Roses

A rose is a woody perennial of the genus *Rosa*, within the family Rosaceae. There are over 100 species. They form a group of plants that can be erect shrubs, climbing or trailing with stems that are often armed with sharp prickles. Flowers vary in size and shape and are usually large and showy, in colours ranging from white through yellows and reds. Most species are native to Asia, with smaller numbers native to Europe, North America, and northwest Africa. Species, cultivars and hybrids are all widely grown for their beauty and often are fragrant. Rose plants range in size from compact, miniature roses, to climbers that can reach 7 metres in height. Different species hybridize easily, and this has been used in the development of the wide range of garden roses.

Botany

The leaves are borne alternately on the stem. In most species they are 5 to 15 centimetres (2.0 to 5.9 in) long, pinnate, with (3–) 5–9 (–13) leaflets and basal stipules; the leaflets usually have a serrated margin, and often a few small prickles on the underside of the stem. Most roses are deciduous but a few (particularly from South east Asia) are evergreen or nearly so.

The flowers of most species have five petals, with the exception of *Rosa sericea*, which usually has only four. Each petal is divided into two distinct lobes and is usually white or pink, though in a few species yellow or red. Beneath the petals are five sepals (or in the case of some *Rosa sericea*, four). These may be long enough to be visible when viewed from above and appear as green points alternating with the rounded petals. There are multiple superior ovaries that develop into achenes. Roses are insect-pollinated in nature.

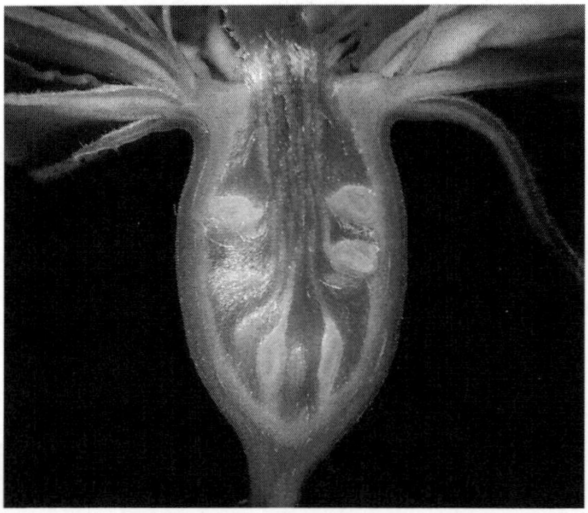

Figure: *Cross-section through a developing rose hip*

The aggregate fruit of the rose is a berry-like structure called a rose hip. Many of the domestic cultivars do not produce hips, as the flowers are so tightly petalled that they do not provide access for pollination. The hips of most species are red, but a few (e.g. *Rosa pimpinellifolia*) have dark purple to black hips. Each hip comprises an outer fleshy layer, the hypanthium, which contains 5–160 "seeds" (technically dry single-seeded fruits called achenes) embedded in a matrix of fine, but stiff, hairs. Rose hips of some species, especially the Dog Rose (*Rosa canina*) and Rugosa Rose (*Rosa rugosa*), are very rich in vitamin C, among the richest sources of any plant. The hips are eaten by fruit-eating birds such as thrushes and waxwings, which then disperse the seeds in their droppings. Some birds, particularly finches, also eat the seeds.

While the sharp objects along a rose stem are commonly called "thorns", they are technically prickles — outgrowths of the epidermis (the outer layer of tissue of the stem). (True thorns, as produced by e.g. *Citrus* or *Pyracantha*, are modified stems, which always originate at a node and which have nodes and internodes along the length of the thorn itself.) Rose prickles are typically sickle-shaped hooks, which aid the rose in hanging onto other vegetation when growing over it. Some species such as *Rosa rugosa* and *Rosa pimpinellifolia* have densely packed straight prickles, probably an adaptation to reduce browsing by animals, but also possibly an adaptation to trap wind-blown sand and so reduce erosion and protect their roots (both of these species grow naturally on coastal sand dunes). Despite the presence of prickles, roses are frequently browsed by deer. A few species of roses have only vestigial prickles that have no points.

Species

***Figure:** Rosa viridifolia, the petals resemnle leaves*

The genus *Rosa* is subdivided into four subgenera:

- *Hulthemia* (formerly *Simplicifoliae*, meaning "with single leaves") containing one or two species from southwest Asia, *R. persica* and *Rosa berberifolia* which are the only roses without compound leaves or stipules.
- *Hesperrhodos* (from the Greek for "western rose") contains *Rosa minutifolia* and *Rosa stellata*, from North America.
- *Platyrhodon* (from the Greek for "flaky rose", referring to flaky bark) with one species from east Asia, *Rosa roxburghii*.
- *Rosa* (the type subgenus, incorrectly called *Eurosa*) containing all the other roses. This subgenus is subdivided into 11 sections.
    - *Banksianae* - white and yellow flowered roses from China.
    - *Bracteatae* - three species, two from China and one from India.
    - *Caninae* - pink and white flowered species from Asia, Europe and North Africa.
    - *Carolinae* - white, pink, and bright pink flowered species all from North America.
    - *Chinensis* - white, pink, yellow, red and mixed-colour roses from China and Burma.
    - *Gallicanae* - pink to crimson and striped flowered roses from western Asia and Europe.
    - *Gymnocarpae* - one species in western North America (*Rosa gymnocarpa*), others in east Asia.

- o *Laevigatae* - a single white flowered species from China
- o *Pimpinellifoliae* - white, pink, bright yellow, mauve and striped roses from Asia and Europe.
- o *Rosa* (syn. sect. *Cinnamomeae*) - white, pink, lilac, mulberry and red roses from everywhere but North Africa.
- o *Synstylae* - white, pink, and crimson flowered roses from all areas.

Uses

Roses are best known as ornamental plants grown for their flowers in the garden and sometimes indoors. They have been also used for commercial perfumery and commercial cut flower crops. Some are used as landscape plants, for hedging and for other utilitarian purposes such as game cover and slope stabilization. They also have minor medicinal uses.

Garden Roses

Garden roses are predominantly hybrid roses that are grown as ornamental plants in private or public gardens. They are one of the most popular and widely cultivated groups of flowering plants, especially in temperate climates. Numerous cultivars have been produced, especially over the last two centuries, though roses have been known in the garden for millennia beforehand. While most garden roses are grown for their flowers, some are also valued for other reasons, such as having ornamental fruit, providing ground cover, or for hedging.

It is believed that roses were grown in all the early civilisations of temperate latitudes from at least 5000 years ago. They are known to have been grown in ancient Babylon. Paintings of roses have been discovered in Egyptian pyramid tombs from the 14th century BC. Records exist of them being grown in Chinese gardens and Greek gardens from at least 500 BC.

Most of the plants grown in these early gardens are likely to have been species collected from the wild. However there were large numbers of selected varieties being grown from early times; for instance numerous selections or cultivars of the China rose were in cultivation in China in the first millennium AD.

The significant breeding of modern times started slowly in Europe, from about the 17th century. This was encouraged by the introduction of new species, and especially by the introduction of the China rose into Europe in the 19th century. An enormous range of roses has been bred since then. A major contributor in the early 19th century was Empress Josephine of France who patronized the development of rose breeding at her gardens at Malmaison. As long ago as 1840 a collection numbering over one thousand different cultivars, varieties and species was possible when a rosarium was planted by Loddiges nursery for Abney Park Cemetery, an early Victorian garden cemetery and arboretum in England.

Classification

There is no single system of classification for garden roses. In general, however, roses are placed in one of three main groups: Wild, Old Garden, and Modern Garden roses. The latter two groups are usually subdivided further according to hybrid lineage, although due to the complex ancestry of most rose hybrids, such distinctions can be imprecise. Growth habit and floral form are also used as means of classification.

Wild Roses

The wild roses (also known as species roses) include the natural species and some of their immediate hybrid descendents. The wild roses commonly grown in gardens include *Rosa moschata*, the Musk Rose; *Rosa banksiae*, Lady Banks' Rose; *Rosa pimpinellifolia*, the Scots or Burnet Rose; *Rosa rubiginosa*, the Sweetbriar or Eglantine; and *Rosa foetida*, in varieties 'Austrian Copper', 'Persian Double' and 'Harison's Yellow'. For most of these, the plants found in cultivation are often selected clones that are propagated vegetatively. Wild roses are low-maintenance shrubs in comparison to other garden roses, usually tolerating poor soil and some shade. They generally have only one flush of blooms per year, although some species have large hips in the autumn (e.g. *Rosa moyesii*) or have colourful autumn foliage (e.g. *Rosa virginiana*)

Old Garden Roses

An Old Garden Rose is defined as any rose belonging to a class which existed before the introduction of the first Modern Rose, La France, in 1867. Alternative terms for this group include heritage and historic roses. In general, Old Garden Roses of European or Mediterranean origin are once-blooming woody shrubs, with notably fragrant, double-flowered blooms primarily in shades of white, pink and crimson-red. The shrubs' foliage tends to be highly disease-resistant, and they generally bloom only from canes (stems) which formed in previous years. The introduction of China and Tea roses from East Asia around 1800 led to new classes of Old Garden Roses which bloom on new growth, often repeatedly from spring to fall. Most Old Garden Roses are classified into one of the following groups.

Alba

Literally "white roses", derived from *R. arvensis* and the closely allied *R. x alba*. The latter species is a hybrid of *R. gallica* and *R. canina*. This group contains some of the oldest garden roses. The shrubs flower once yearly in the spring or early summer with scented blossoms of white or pale pink. They frequently have gray-green foliage and a vigorous or climbing habit of growth. Examples are 'Alba Semiplena', 'White Rose of York'.

Gallica

The Gallica or Rose of Provins group is a very old class developed from *Rosa gallica*, which is a native of central and southern Europe and western Asia. The Apothecary's Rose, *R. gallica officinalis,* was grown in the Middle Ages in monastic *herbaria* for its alleged medicinal properties, and became famous in English history as the Red Rose of Lancaster. Gallicas flower once in the summer over low shrubs rarely over 4 feet (1.25m) tall. Unlike most other once-blooming Old Garden Roses, the gallica class includes shades of red, maroon and deep purplish crimson. Examples are 'Cardinal de Richelieu', 'Charles de Mills', 'Rosa Mundi' (*R. gallica versicolour*).

Damask

Named for Damascus, Damask roses (*Rosa x damascena*) originated in ancient times with a natural cross (*Rosa moschata* x *Rosa gallica*) x *Rosa fedtschenkoana.* Robert de Brie is given credit for bringing damask roses from the Middle East to Europe sometime between 1254 and 1276, although there is evidence from ancient Roman frescoes that at least one damask rose existed in Europe for hundreds of years before this. Summer damasks bloom once in summer. Autumn or Four Seasons damasks bloom again later, albeit less exuberantly, and these were the first remontant (repeat-flowering) Old European roses. Damask roses tend to have rangy to sprawling growth habits and strongly scented blooms. Examples: 'Ispahan', 'Madame Hardy'.

Centifolia or Provence

Centifolia roses are also known as Cabbage roses, or as Provence roses. They are derived from *Rosa* x *centifolia,* a hybrid that appeared in the 17th century in the Netherlands, related to damask roses. They are named for their "one hundred" petals; they are often called "cabbage" roses due to the globular shape of the flowers. The centifolias are all once-flowering. As a class, they are notable for their inclination to produce mutations of various sizes and forms, including moss roses and some of the first miniature roses. Examples: 'Centifolia', 'Paul Ricault'.

Moss

The Moss roses are based on one or more mutations, particularly one that appeared early on *Rosa centifolia*, the Provence or cabbage rose. Some with Damask roses as a parent may be derived from a separate mutation. Thickly growing or branched resin-bearing hairs, particularly on the sepals, give off a pleasant woods or balsam scent when rubbed. Moss roses are cherished for this trait, but as a group they have not contributed to the development of new rose classifications. Various hybrids with other roses have yielded different forms, such as the modern miniature creeping moss rose 'Red Moss Rambler' (Ralph S. Moore, 1990). Moss roses with centifolia background are once-flowering; some moss roses exhibit repeat-blooming, indicative of Autumn Damask parentage. Examples: 'Common Moss' (centifolia-moss), 'Mousseline', also known as 'Alfred de Dalmas' (Autumn Damask moss).

Portland

The Portland roses were long thought to be the first group of crosses between China roses and European roses, and to show the influence of *Rosa chinensis*. Recent DNA analysis however has demonstrated that the original Portland Rose has no Chinese ancestry, but has an autumn damask/gallica lineage. This group of roses was named after the Duchess of Portland who received (from Italy about 1775) a rose then known as *R. paestana* or 'Scarlet Four Seasons' Rose' (now known simply as 'The Portland Rose'). The whole class of Portland roses was developed from that one rose. The first repeat-flowering class of rose with fancy European-style blossoms, the plants tend to be fairly short and shrubby, with a suckering habit, with proportionately short flower stalks. The main flowering is in the summer, but intermittent flowers continue into the autumn. Examples: 'James Veitch', 'Rose de Rescht', 'Comte de Chambord'.

China

The China roses, based on *Rosa chinensis*, have been cultivated in East Asia for centuries. They have been cultivated in Western Europe since the late 18th century. They contribute much to the parentage of today's hybrid roses, and they brought a change to the form of the flowers then cultivated in Europe. Compared with the older rose classes known in Europe, the Chinese roses had less fragrant, smaller blooms carried over twiggier, more cold-sensitive shrubs. However they could bloom repeatedly throughout the summer and into late autumn, unlike their European counterparts. The flowers of China roses were also notable for their tendency to "suntan," or darken over timeunlike other blooms which tended to fade after opening. This made them highly desirable for hybridisation purposes in the early 19th century. According to Graham Stuart Thomas, China Roses are the class upon which modern roses are built. Today's exhibition rose owes its form to the China genes, and the China Roses also brought slender buds which unfurl when opening. Tradition holds that four "stud China" roses—'Slater's Crimson China' (1792), 'Parsons' Pink China' (1793), and the Tea roses 'Hume's Blush Tea-scented China' (1809) and 'Parks' Yellow Tea-scented China' (1824)—were brought to Europe in the late 18th and early 19th centuries; in fact there were rather more, at least five Chinas not counting the Teas having been imported. This brought about the creation of the first classes of repeat-flowering Old Garden Roses, and later the Modern Garden Roses. Examples: 'Old Blush China', 'Mutabilis' (Butterfly Rose), 'Cramoisie Superieur'.

Tea

The original Tea-scented Chinas (*Rosa* × *odorata*) were Oriental cultivars thought to represent hybrids of *R. chinensis* with *R. gigantea*, a large Asian climbing rose with pale-yellow blossoms. Immediately upon their introduction in the early 19th-century breeders went to work with them, especially in France, crossing them first with China roses and then with Bourbons and Noisettes. The Tea roses are repeat-flowering roses, named for their fragrance being reminiscent of Chinese black tea (although this is not always the case). The colour range includes pastel shades of white, pink and (a novelty at the time) yellow to apricot. The individual flowers of many cultivars are semi-pendent and nodding, due to weak flower stalks. In a "typical" Tea, pointed buds produce high-centred blooms which unfurl in a spiral fashion, and the petals tend to roll back at the edges, producing a petal with a pointed tip; the Teas are thus the originators of today's "classic" florists' rose form. According to rose historian Brent Dickerson, the Tea classification owes as much to marketing as to botany; 19th-century nurserymen would label their Asian-based cultivars as "Teas" if they possessed the desirable Tea flower form, and "Chinas" if they did not. Like the Chinas, the Teas are not hardy in colder climates. Examples: 'Lady Hillingdon', 'Maman Cochet', 'Duchesse de Brabant', 'Mrs. Foley Hobbs'.

Bourbon

Bourbon roses originated on the Île Bourbon (now called Réunion) off the coast of Madagascar in the Indian Ocean. They are believed to be the result of a cross between the Autumn Damask and the 'Old Blush' China rose, both of which were frequently used as hedging materials on the island. They flower repeatedly on vigorous, frequently semi-climbing shrubs with glossy foliage and purple-tinted canes. They were first Introduced in France in 1823 by Henri Antoine Jacques. Examples: 'Louise Odier', 'Mme. Pierre Oger', 'Zéphirine Drouhin' (the last example is often classified under climbing roses).

Noisette

The first Noisette rose was raised as a hybrid seedling by a South Carolina rice planter named John Champneys. Its parents were the China Rose 'Parson's Pink' and the autumn-flowering musk rose (*Rosa moschata*), resulting in a vigorous climbing rose producing huge clusters of small pink flowers from spring to fall. Champneys sent seedlings of his rose (called 'Champneys' Pink Cluster') to his gardening friend, Philippe Noisette, who in turn sent plants to his brother Louis in Paris, who then introduced 'Blush Noisette' in 1817. The first Noisettes were small-blossomed, fairly winter-hardy climbers, but later infusions of Tea rose genes created a Tea-Noisette subclass with larger flowers, smaller clusters, and considerably reduced winter hardiness. Examples: 'Blush Noisette', 'Lamarque' (Noisette); 'Mme. Alfred Carriere', 'Marechal Niel' (Tea-Noisette).

Hybrid Perpetual

The dominant class of roses in Victorian England, hybrid perpetuals (a misleading translation of *hybrides remontants*, 'reblooming hybrids') emerged in 1838 as the first roses which successfully combined Asian remontancy (repeat blooming) with the old European lineages. Since re-bloom is a recessive trait, the first generation of Asian/European crosses (Hybrid Chinas, Hybrid Bourbons, Hybrid Noisettes) were stubbornly once-blooming, but when these roses were recrossed with themselves or with Chinas or teas, some of their offspring flowered more than once. The Hybrid Perpetuals thus were something of a miscellany, a catch-all class derived to a great extent from the Bourbons but with admixtures of Chinas, teas, damasks, gallicas, and to a lesser extent Noisettes, albas and even centifolias. They became the most popular garden and florist roses of northern Europe at the time, as the tender tea roses would not thrive in cold climates, and the Hybrid Perpetuals' very large blooms were well-suited to the new phenomenon of competitive exhibitions. The "perpetual" in the name hints at repeat-flowering, but many varieties of this class had poor re-flowering habits; the tendency was for a massive spring bloom, followed by either scattered summer flowering, a smaller autumn burst, or sometimes nothing at all until next spring. Due to a limited colour palette (white, pink, red) and lack of reliable repeat-bloom, the hybrid perpetuals were ultimately overshadowed by their descendants, the Hybrid Teas. Examples: 'Ferdinand Pichard', 'Reine Des Violettes', 'Paul Neyron'.

Hybrid Musk

Although they arose too late to qualify technically as Old Garden Roses, the hybrid musks are often informally classed with them, since their growth habits and care are much more like the Old Garden Roses than Modern Roses. The hybrid musk group was mainly developed by Rev. Joseph Pemberton, a British rosarian, in the first decades of the 20th century, based upon 'Aglaia', an 1896 cross by Peter Lambert. A seedling of this rose, 'Trier', is considered to the foundation of the class. The genetics of the class are somewhat obscure, as some of the parents are unknown. *Rose multiflora*, however, is known to be one parent, and *Rosa moschata* (the musk rose) also figures in its heritage, though it is considered to be less important than the name would suggest. Hybrid musks are disease-resistant, repeat flowering and generally cluster-flowered, with a strong, characteristic "musk" scent. The stems tend to be lax and arching, with limited thorns. Examples include 'Buff Beauty' and 'Penelope'.

Hybrid Rugosa

Figure: *Rugosa rose 'Blanc Double de Coubert' (Cochet 1893)*

The Rugosas likewise are not officially Old Garden Roses, but tend to be grouped with them. Derived from *Rosa rugosa* from Japan and Korea beginning in the 1880s, these vigorous roses are extremely hardy with excellent disease resistance. Most are extremely fragrant, repeat bloomers with moderately double flat flowers. The defining characteristic of a Hybrid Rugosa rose is its wrinkly leaves, but some hybrids do lack this trait. These roses will often set hips. Examples include 'Hansa' and 'Roseraie de l'Häy'.

Bermuda "Mystery" Roses

This is a group of several dozen "found" roses grown in Bermuda for at least a century. The roses have significant value and interest for those growing roses in tropical and semi-tropical regions, since they are highly resistant to both nematode damage and the fungal diseases that plague rose culture in hot, humid areas. They are also capable of thriving in hot and humid weather. Most of these roses are thought to be Old Garden Rose cultivars that have otherwise dropped out of cultivation, or sports thereof. They are "mystery roses" because their "proper" historical names have been lost. Tradition dictates that they are named after the owner of the garden where they were rediscovered.

Miscellaneous

There are also a few smaller classes (such as Scots, Sweet Brier) and some climbing classes of old roses (including Ayrshire, Climbing China, Laevigata, Sempervirens, Boursault, Climbing Tea, and Climbing Bourbon). Those classes with both climbing and shrub forms are often grouped together.

Modern Garden Roses

Classification of modern roses can be quite confusing because many modern roses have old garden roses in their ancestry and their form varies so much. The classifications tend to be by growth and flowering characteristics. The following includes the most notable and popular classifications of Modern Garden Roses:

Hybrid Tea

The favourite rose for much of the history of modern roses, hybrid teas were initially created by hybridising Hybrid Perpetuals with Tea roses in the late 19th century. 'La France', created in 1867, is universally acknowledged as the first indication of a new class of roses. Hybrid teas exhibit traits midway between both parents: hardier than the teas but less hardy than the hybrid perpetuals, and more ever-blooming than the hybrid perpetuals but less so than the teas. The flowers are well-formed with large, high-centred buds, and each flowering stem typically terminates in a single shapely bloom. The shrubs tend to be stiffly upright and sparsely foliaged, which today is often seen as a liability because it makes them more difficult to place in the garden or landscape. Hybrid teas became the single most popular garden rose of the 20th century; today, their reputation as high maintenance plants has led to a decline in popularity. The hybrid tea remains the standard rose of the floral industry, however, and is still favoured in formal situations. Examples: 'Peace' (yellow), 'Garden Party' (white), 'Mister Lincoln' (red) and 'Double Delight' (bi-colour cream and red).

Pernetiana

The French breeder Joseph Pernet-Ducher initiated the first class of roses to include genes from the old Austrian briar rose (*Rosa foetida*) with his 1900 introduction of 'Soleil d'Or.' This resulted in an entirely new colour range for roses: shades of deep yellow, apricot, copper, orange, true scarlet, yellow bicolours, lavender, gray, and even brown were now possible. Originally considered a separate class, the Pernetianas or Hybrid Foetidas were officially merged into the Hybrid Teas in 1930. The new colour range did much to increase hybrid tea popularity in the 20th century, but these colours came at a price: *Rosa foetida* also passed on a tendency towards disease-susceptibility, scentless blooms, and an intolerance of pruning to its descendants.

Polyantha

Literally "many-flowered" roses, from the Greek "poly" (many) and "anthos" (flower). Originally derived from crosses between two East Asian species (*Rosa chinensis* and *Rosa multiflora*), polyanthas first appeared in France in the late 19th century alongside the hybrid teas. They featured short plants—some compact, others spreading in habit—with tiny blooms (2.5 cm or 1 inch in diametre on average) carried in large sprays, in the typical rose colours of white, pink and red. Their main claim to fame was their prolific bloom: From spring to fall, a healthy polyantha shrub might be literally covered in flowers, creating a strong colour impact in the landscape. Polyantha roses are still regarded as low-maintenance, disease-resistant garden roses today, and remain popular for that reason. Examples: 'Cécile Brünner', 'The Fairy', 'Red Fairy', 'Pink Fairy'.

Floribunda

Some rose breeders saw potential in crossing polyanthas with hybrid teas, to create roses that bloomed with the polyantha profusion, but with hybrid tea floral beauty and colour range. In 1907, the first polyantha/hybrid tea cross, 'Rödhätte', was introduced by Danish breeder Dines Poulsen. This had some characteristics of both parent classes, and was termed a Hybrid Polyantha or Poulsen rose. Further similar introductions followed from Poulsen, these often bearing the family name e.g. 'Else Poulsen' (1924). As their hybrid characteristics separated these new roses from polyanthas and hybrid teas alike, a new class was created and named *floribunda*, Latin for "many-flowering." Typical floribundas feature stiff shrubs, smaller and bushier than the average hybrid tea but less dense and sprawling than the average polyantha. The flowers are often smaller than hybrid teas but are carried in large sprays, giving a better floral effect in the garden. Floribundas are found in all hybrid tea colours and often with the classic hybrid tea-shaped blossom, sometimes differing from hybrid teas only in their cluster-flowering habit. Today they are still used in large bedding schemes in public parks and similar spaces. Examples: 'Anne Harkness', 'Dainty Maid', 'Iceberg', 'Tuscan Sun'.

Grandiflora

Grandifloras (Latin for "large-flowered") were the class of roses created in the mid-20th century to designate back-crosses between hybrid teas and floribundas that fit neither category—specifically, the 'Queen Elizabeth' rose, which was introduced in 1954. Grandiflora shrubs are typically larger than either hybrid teas or floribundas, and feature hybrid tea-style flowers borne in small clusters of three to five, similar to a floribunda. Grandifloras maintained some popularity from about the 1950s to the 1980s but today they are much less popular than either the hybrid teas or the floribundas. Examples: 'Queen Elizabeth', 'Comanche,' 'Montezuma'.

Miniature

Dwarf mutations of some Old Garden Roses—gallicas and centifolias—were known in Europe in the 17th century, although these were once-flowering just as their larger forms were. Miniature forms of repeat-flowering China roses were also grown and bred in China, and are depicted in 18th-century Chinese art. Modern miniature roses largely derive from such miniature China roses, especially the cultivar 'Roulettii', a chance discovery found in a pot in Switzerland.

Miniature roses are represented by twiggy, repeat-flowering shrubs ranging from 6" to 36" in height, with most falling in the 12"–24" height range. Blooms come in all the hybrid tea colours; many varieties also emulate the classic high centred hybrid tea flower shape. Miniature roses are often marketed and sold by the floral industry as houseplants, but it is important to remember that these plants are largely descended from outdoor shrubs native to temperate regions; thus, most miniature rose varieties require an annual period of cold dormancy to survive. (Examples: 'Petite de Hollande' (Miniature Centifolia, once-blooming), 'Cupcake' (Modern Miniature, repeat-blooming). Miniature garden roses only grow in the summer.

Climbing and Rambling

All aforementioned classes of roses, both Old and Modern, have "climbing" forms, whereby the canes of the shrubs grow much longer and more flexible than the normal ("bush") forms. In the Old Garden Roses, this is often simply the natural growth habit; in many Modern roses, however, climbing roses are the results of spontaneous mutations. For example, 'Climbing Peace' is designated as a "Climbing Hybrid Tea," for it is genetically identical to the normal "shrub" form of the 'Peace' hybrid tea rose, except that its canes are long and flexible, i.e. "climbing." Most Climbing roses grow anywhere from 8'–20' in height and exhibit repeat-bloom. Rambler roses, although technically a separate class, are often lumped together with climbing roses. They also exhibit long, flexible canes, but are usually distinguished from true climbers in two ways: A larger overall size (20'–30' tall is common), so is a once-blooming habit. Climbing and rambling roses are not true vines such as ivy, clematis or wisteria because they lack the ability to cling to supports on their own and must be manually trained and tied over structures such as arbors and pergolas. Examples: 'Blaze' (repeat-blooming climber), 'American Pillar' (once-blooming rambler). One of the most vigorous climbing roses is the Kiftsgate rose, *Rosa filipes* 'Kiftsgate', named after the house garden where it was noticed by Graham Stuart Thomas in 1951. The original plant is claimed to be the largest rose in the UK, and has climbed 50 feet into a copper beech tree.

Shrub

This is not a precisely defined class of garden rose, but it is a description or grouping commonly used by rose reference books and catalogues. It encompasses some old single and repeat flowering cultivars, as well as modern roses that don't fit neatly into other categories. Many cultivars placed in other categories are simultaneously placed in this one. Roses classed as shrubs tend to be robust and of informal habit, making them recommended for use in a mixed shrub border or as hedging.

English/David Austin

Although not officially recognized as a separate class of roses by any established rose authority, English (aka David Austin) roses are often set aside as such by consumers and retailers alike. Development started in the 1960s by David Austin of Shropshire, England, who wanted to rekindle interest in Old Garden Roses by hybridizing them with modern hybrid teas and floribundas. The idea was to create a new group of roses that featured blooms with old-fashioned shapes and fragrances, evocative of classic *gallica*, *alba* and "damask" roses, but with modern repeat-blooming characteristics and the larger modern colour range as well. Austin mostly succeeded in his mission; his tribe of "English" roses, now numbering hundreds of varieties, has been warmly embraced by the gardening public and are widely available to consumers. David Austin roses are still actively developed, with new varieties released regularly. The typical winter-hardiness and disease-resistance of the classic Old Garden Roses has largely been compromised in the process; many English roses are susceptible to the same disease problems that plague modern hybrid teas and floribundas, and many are not hardy north of USDA Zone 5. Examples: 'Mary Rose', 'Graham Thomas', 'Tamora'.

Canadian Hardy

Two main lines of roses were developed for the extreme weather conditions of Canadian winters by Agriculture Canada at the Morden Research Station in Morden, Manitoba and the Experimental Farm in Ottawa (and later at L'Assomption, Québec). They are called the Explorer series and the Parkland series. These programmes have now been discontinued, the remaining plant stock has been taken over by private breeders and marketed along with the Canadian Artists roses as a single series. Derived mostly from crosses of *Rosa rugosa* or the native Canadian species *Rosa arkansana* with other species, these plants are extremely tolerant of cold weather, some down to –35C. All have repeat bloom. A wide diversity of forms and colours were achieved.

Examples of roses in the Parkland series include 'Morden Centennial', 'Morden Sunrise, 'Winnipeg Parks' and 'Cuthbert Grant'. Two roses named after Canadian artists that have been added are 'Emily Carr' and 'Felix Leclerc'.

Other notable Canadian breeders include Frank Skinner, Percy Wright, Isabella Preston, Georges Bugnet and Robert Erskine.

Landscape (Ground Cover)

This type of rose was developed mainly for mass amenity planting. In the late 20th century, traditional hybrid tea and floribunda rose varieties fell out of favour with many gardeners and landscapers, as they are often labour and chemical intensive plants susceptible to pest and disease problems. So-called "landscape" roses (also known as "ground cover" roses) have thus been developed to fill the consumer desire for a garden rose that offers colour, form and fragrance, but is also low maintenance and easy to care for. Most have the following characteristics:

- Lower growing habit, usually under 60 cm (24 inches)
- Repeat flowering
- Disease and pest resistance
- Growing on their own roots.
- Minimal pruning requirements

Principal parties involved in the breeding of new varieties include: Werner Noak (Germany), Meidiland Roses (France), Boot & Co. (Netherlands), and William Radler (USA).

Patio

Since the 1970s many rose breeders have focused on developing compact roses (typically 1'-3' in height and spread) that are suitable for smaller gardens, terraces and containers. These combine characteristics of larger miniature roses and smaller floribundas—resulting in the rather loose classification "patio roses". Dr. D.G. Hessayon says the description "patio roses" emerged after 1996. Some rose catalogues include older polyanthas that have stood the test of time (e.g., 'Nathalie Nypels', 'Baby Faurax') within their patio selection. Rose breeders, notably Chris Warner in the UK and the Danish firm of Poulson (under the name of Courtyard Climbers) have also created patio climbers, small rambler style plants that flower top-to-toe and are suitable for confined areas.

Cultivation

In the garden, roses are grown as bushes, shrubs or climbers. "Bushes" are usually comparatively low growing, often quite upright in habit, with multiple stems emerging near ground level; they are often grown formally in beds with other roses. "Shrubs" are usually larger and have a more informal or arching habit, and may additionally be placed in a mixed border or grown separately as specimens. Certain bush hybrids (and smaller shrubs) may also be grown as "standards", which are plants grafted high (typically 1 metre or more) on a rose rootstock, resulting in extra height which can make a dominant feature in a floral display. Climbing roses are usually trained to a suitable support.

Roses are commonly propagated by grafting onto a rootstock, which provides sturdiness and vigour, or (especially with Old Garden Roses) they may be propagated from hardwood cuttings and allowed to develop their own roots.

Most roses thrive in temperate climates. Those based on warm climate Asian species do well in their native sub-tropical environments.

Certain species and cultivars can even flourish in tropical climates, especially when grafted onto appropriate rootstocks. Most garden roses prefer rich soil which is well-watered but well-drained, and perform best in well-lit positions which receive several hours of sun a day (although some climbers, some species and most Hybrid Musks will tolerate shade). Standard roses require staking.

Pruning

Rose pruning, sometimes regarded as a horticultural art form, is largely dependent on the type of rose to be pruned, the reason for pruning, and the time of year it is at the time of the desired pruning. Most Old Garden Roses of strict European heritage (albas, damasks, gallicas, etc.) are shrubs that bloom once yearly, in late spring or early summer, on two-year-old (or older) canes. Their pruning requirements are quite minimal because removal of branches will remove next year's flower buds. Hence pruning is usually restricted to just removing weak and spent branches, plus light trimming (if necessary) to reduce overall size.

Modern hybrids, including the hybrid teas, floribundas, grandifloras, modern miniatures, and English roses, have a complex genetic background that almost always includes China roses (which are descended from *Rosa chinensis*). China roses were evergrowing, everblooming roses from humid subtropical regions that bloomed constantly on any new vegetative growth produced during the growing season. Their modern hybrid descendants exhibit similar habits; unlike Old European Roses, modern hybrids bloom continuously (until stopped by frost) on any new canes produced during the growing season. They therefore require pruning back of any spent flowering stem in order to divert the plant's energy into producing new growth and hence new flowers.

Additionally, Modern Hybrids planted in cold winter climates will almost universally require a "hard" annual pruning (reducing all canes to 8"–12", about 30 cm in height) in early spring. Again, because of their complex China rose background, modern hybrids are typically not as cold hardy as European Old Garden Roses, and low winter temperatures often desiccate or kill exposed canes. In spring, if left unpruned, these damaged canes will often die back all the way to the shrub's root zone, resulting in a weakened, disfigured plant. The annual "hard" pruning of hybrid teas and floribundas is generally done in early spring.

Deadheading

This is the practice of removing any spent, faded, withered, or discoloured flowers. The purpose is to encourage the plant to focus its energy and resources on forming new shoots and blooms, rather than fruit production. Deadheading may also be performed for aesthetic purposes, if spent flowers are unsightly. Any roses such as Rosa glauca or Rosa moyesii that are grown for their decorative hips should not be deadheaded.

Pests and Diseases

Roses are subject to several diseases. The main fungal diseases affecting the leaves are rose black spot (*Diplocarpon rosae*), rose rust (*Phragmidium mucronatum*), rose powdery mildew (*Sphaerotheca pannosa*) and rose downy mildew (*Peronospora sparsa*). Stems can be affected by several canker diseases, the most commonly seen of which is stem canker (*Leptosphaeria coniothyrium*). Diseases of the root zone include honey fungus (*Armillaria* spp.), verticillium wilt, and various species of *phytophthora*. Fungal leaf diseases affect some cultivars and species more than others. On susceptible plants fungicidal sprays may be necessary to prevent infection or reduce severity of attacks. Cultivation techniques may also be used, such as ensuring good air circulation around a plant. Stem cankers are best treated by pruning out infection as soon as it is noticed. Root diseases are not usually possible to treat once infection has occurred; the most practical line of defence is to ensure that growing conditions maximise plant health and thereby prevent infection. *Phytophthora* species are waterborne and therefore improving drainage and reducing waterlogging can help reduce infection.

The main pest affecting roses is the aphid (greenfly), which sucks the sap and weakens the plant. In areas where they are endemic Japanese beetles (*Popillia japonica*) take a heavy toll on rose flowers and foliage; rose blooms can also be destroyed by infestations of thrips (*Thysanoptera* spp). Roses are also used as food plants by the larvae of some Lepidoptera (butterfly and moth) species. Spraying with insecticide of roses is often recommended but if this is done care is needed to minimize the loss of beneficial insects; systemic insecticides have the advantage of only affecting insects which feed on the plants.

Breeding and Commercial Growers

Tyler, Texas, has been nicknamed the "Rose Capital of America" because of its large role in the rose-growing industry. It boasts the nation's largest municipal rose garden and hosts the Texas Rose Festival each October, which draws thousands of spectators.

Perfume

Rose perfumes are made from attar of roses or rose oil, which is a mixture of volatile essential oils obtained by steam distilling the crushed petals of roses. An associated product is rose water which is used for cooking, cosmetics, medicine and in religious practices. The production technique originated in Persia then spread through Arabia and India, and more recently into eastern Europe. In Bulgaria, Iran and Germany, damask roses (*Rosa* × *damascena* 'Trigintipetala') are used. In other parts of the world *Rosa* × *centifolia* is commonly used. The oil is transparent pale yellow or yellow-grey in colour. 'Rose Absolute' is solvent-extracted with hexane and produces a darker oil, dark yellow to orange in colour. The weight of oil extracted is about one three-thousandth to one six-thousandth of the weight of the flowers; for example, about two thousand flowers are required to produce one gram of oil.

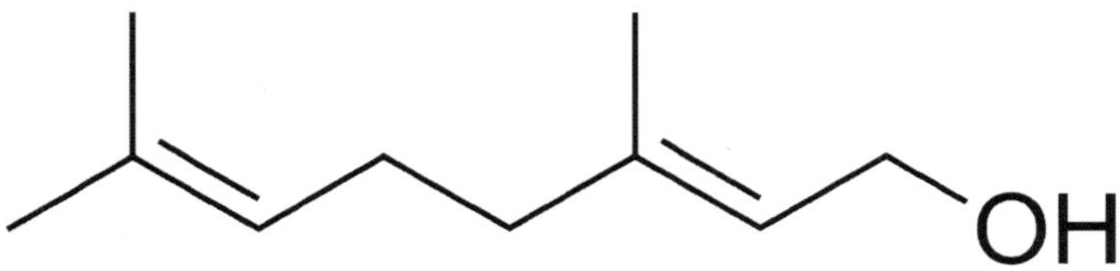

Figure: *Geraniol* ($C_{10}H_{18}O$)

The main constituents of attar of roses are the fragrant alcohols geraniol and l-citronellol and rose camphor, an odorless solid composed of alkanes, which separates from rose oil. β-Damascenone is also a significant contributor to the scent.

Food and Drink

Rose hips are occasionally made into jam, jelly, and marmalade, or are brewed for tea, primarily for their high vitamin C content. They are also pressed and filtered to make rose hip syrup. Rose hips are also used to produce Rose hip seed oil, which is used in skin products and some makeup products. Rose petals or flower buds are sometimes used to flavour ordinary tea, or combined with other herbs to make herbal teas. In France there is much use of rose syrup, most commonly made from an extract of rose petals. In the United States, this French rose syrup is used to make rose scones and marshmallows.

Rose flowers are used as food, also usually as flavouring or to add their scent to food. Other minor uses include candied rose petals. Rose creams (rose flavoured fondant covered in chocolate, often topped with a crystallised rose petal) are a traditional English confectionery widely available from numerous producers in the UK.

Medicine

The rose hip, usually from *R. canina* is used as a minor source of Vitamin C. The fruits of many species have significant levels of vitamins and have been used as a food supplement. Many roses have been used in herbal and folk medicines. *Rosa chinensis* has long been used in Chinese traditional medicine. This and other species have been used for stomach problems, and are being investigated for controlling cancer growth.

Symbolism

The long cultural history of the rose has led to it being used often as a symbol.

Piedmont Azalea

Azaleas are flowering shrubs comprising two of the eight subgenera of the genus *Rhododendron*: the *Tsutsuji* (evergreen) and *Pentanthera* (deciduous). Azaleas bloom in spring, their flowers often lasting several weeks. Shade tolerant, they prefer living near or under trees.

Piedmont Azalea is definitely one of the loveliest blooming shrubs in the South. And the fact that it is a native plant means it should be planted in every southern garden.

The Piedmont Azalea, Rhododendron canescens, is native to the Piedmont region of the Southeast, making it suitable for growing anywhere in Georgia.

The fragrant blooms can be anywhere from a vivid pink to a soft pink or even a pinkish white. Blooms appear in very early spring before the leaves on this deciduous shrub.

Although Piedmont Azalea naturally occurs in the warm southern United States, it is hardy to areas as cold as USDA Zone 5.

All azaleas prefer well-drained soil, so you should amend your soil with soil conditioner or composted bark to improve drainage if your soil is clay.

Native Azaleas are usually found growing in the woods but will bloom more profusely in full sun as long as water is adequate.

Bloom buds are formed in late summer and early fall, so pay close attention to watering during this time. When a native azalea fails to bloom, lack of water during bud formation is usually the culprit.

Native azaleas are drought tolerant once established, but water weekly the first year or two to make sure your plant gets a good start.

Cultivation

Plant enthusiasts have selectively bred azaleas for hundreds of years. This human selection has produced over 10,000 different cultivars which are propagated by cuttings. Azalea seeds can also be collected and germinated. Azaleas are generally slow-growing and do best in well-drained acidic soil (4.5–6.0 pH). Fertilizer needs are low; some species need regular pruning.

Azaleas are native to several continents including Asia, Europe and North America. They are planted abundantly as ornamentals in the southeastern USA, southern Asia,and parts of southwest Europe.

According to azalea historian Fred Galle, in the United States, *Azalea indica* (in this case, the group of plants called Southern indicas) was first introduced to the outdoor landscape in the 1830s at the rice plantation Magnolia-on-the-Ashley in Charleston, South Carolina. Magnolia's owner John Grimke Drayton imported the plants for use in his estate garden from Philadelphia, where they were grown only in greenhouses. With encouragement from Charles Sprague Sargent from Harvard's Arnold Arboretum, Magnolia Gardens was opened to the public in 1871, following the American Civil War. Magnolia is one of the oldest public gardens in America.

Disease

Azalea leafy gall can be particularly destructive to azalea leaves during the early spring. Hand picking infected leaves is the recommended method of control.

They can also be subject to phytophthora root rot in moist, hot conditions.

Cultural Significance and Symbolism

In Chinese culture, the azalea is known as "thinking of home bush" (sixiang shu) and is immortalized in the poetry of Du Fu and is used to rich effect in contemporary stories such as by Taiwanese author Pai, Hsien-Yung. The azalea is also one of the symbols of the city of São Paulo, in Brazil.

In addition to being renowned for its beauty, the Azalea is also highly toxic—it contains andromedotoxins in both its leaves and nectar, including honey from the nectar. The Azalea and Rhododendron were once so infamous for their toxicity that to receive a bouquet of their flowers in a black vase was a well-known death threat.

Azalea Festivals

Japan: Motoyama, Kochi also has a flower festival in which the blooming of Tsutsuji is celebrated and Tatebayashi, Gunma is famous for its Azalea Hill Park, Tsutsuji-ga-oka. Nezu Shrine in Bunkyo, Tokyo, holds a Tsutsuji Matsuri from early April until early May.

Korea

Sobaeksan, one of the 12 well-known Sobaek Mountains, lying on the border between Chungbuk Province and Gyeongbuk has a Royal Azalea (*Rhododendron schlippenbachii*) Festival held on May every year. Sobaeksan has an azalea colony dotted around Biro mountaintop, Gukmang and Yonwha early in May. When Royal azaleas have turned pink in the end of May, it looks like Sobaeksan wears a pink Jeogori (Korean traditional jacket). (Information from Dpt. of Culture & Tourism, Danyang-gun County Office)

United States

Many cities in the United States have festivals in the spring celebrating the blooms of the azalea, including Summerville, South Carolina; Hamilton, NJ; Mobile, Alabama; Jasper, Texas; Norfolk, Virginia; Wilmington, North Carolina (North Carolina Azalea Festival); Valdosta, Georgia; Palatka, Florida (Florida Azalea Festival); Pickens, South Carolina; Muskogee, Oklahoma; Brookings, Oregon; and Stollysville, Ohio. The Azalea Trail is a designated path, planted with azaleas in private gardens, through Mobile, Alabama. The Azalea Trail Run is an annual road running event held there in late March. Mobile, Alabama is also home to the Azalea Trail Maids, fifty women chosen to serve as ambassadors of the city while wearing antebellum dresses, who originally participated in a three-day festival, but now operate throughout the year.

Hong Kong

This is Ma On Shan Azalea Festival in Ma On Shan, where six native species (*Rhododendron championae, Rhododendron farrerae, Rhododendron hongkongense, Rhododendron moulmainense, Rhododendron simiarum* and *Rhododendron simsii*) are found in the area. The Festival has been held since 2004;many activities, such as, exhibitions, photo contests and carnivals are held during the festival.

Gelsemium Sempervirens

Gelsemium sempervirens is a twining vine in the family Gelsemiaceae, native to warm temperate and tropical America from Guatemala north to the Southeastern United States. It has a number of common names including yellow jessamine or jasmine, Carolina jasmine or jessamine, evening trumpetflower, gelsemium and woodbine.

Growth

It can grow to 3–6 m high when given suitable climbing support in trees, with thin stems. The leaves are evergreen, lanceolate, 5–10 cm long and 1-1.5 cm broad, and lustrous, dark green. The flowers are borne in clusters, the individual flowers yellow, sometimes with an orange centre, trumpet-shaped, 3 cm long and 2.5–3 cm broad. Its flowers are strongly scented and produce nectar that attracts a range of pollinators.

Medical Use

Historically *Gelsemium sempervirens* was used as a topical to treat papulous eruptions. It was also used to treat measles, neuralgic otalgia, tonsillitis, esophagitis, dysmenorrhea, muscular rheumatism, headaches.

Toxicity

All parts of this plant contain the toxic strychnine-related alkaloids gelsemine and gelseminine and should not be consumed. The sap may cause skin irritation in sensitive individuals. Children, mistaking this flower for honeysuckle, have been poisoned by sucking the nectar from the flower. The nectar is also toxic to honeybees, which may cause brood death when gathered by the bees. The nectar may, however, be beneficial to bumblebees. It has been shown that bees fed on gelsemine have a reduced load of Crithidia bombi in their fecal mater. Reduced parasite load increases foraging efficiency, and pollinators may selectively collect otherwise toxic secondary metabolites as a means of self-medication.

Despite the hazards, this is a popular garden plant in warmer areas, frequently being trained to grow over arbors or to cover walls.

Yellow Jessamine is the state flower of South Carolina.

Oleander

Nerium oleander is an evergreen shrub or small tree in the dogbane family Apocynaceae, toxic in all its parts. It is the only species currently classified in the genus *Nerium*. It is most commonly known as oleander, from its superficial resemblance to the unrelated olive *Olea*. It is so widely cultivated that no precise region of origin has been identified, though southwest Asia has been suggested. The ancient city of Volubilis in Morocco may have taken its name from the Berber name *oualilt* for the flower. Oleander is one of the most poisonous of commonly grown garden plants.

Oleander grows to 2–6 m (6.6–19.7 ft) tall, with erect stems that splay outward as they mature; first-year stems have a glaucous bloom, while mature stems have a grayish bark. The leaves are in pairs or whorls of three, thick and leathery, dark-green, narrow lanceolate, 5–21 cm (2.0–8.3 in) long and 1–3.5 cm (0.39–1.38 in) broad, and with an entire margin. The flowers grow in clusters at the end of each branch; they are white, pink to red, 2.5–5 cm (0.98–1.97 in) diametre, with a deeply 5-lobed fringed corolla round the central corolla tube. They are often, but not always, sweet-scented. The fruit is a long narrow capsule 5–23 cm (2.0–9.1 in) long, which splits open at maturity to release numerous downy seeds.

Habitat and Range

N. oleander is either native or naturalized to a broad area from Mauritania, Morocco, and Portugal eastward through the Mediterranean region and the Sahara (where it is only found sporadically), to the Arabian peninsula, southern Asia, and as far East as Yunnan in southern parts of China. It typically occurs around dry stream beds. *Nerium oleander* is planted in many subtropical and tropical areas of the world. On the East Coast of the US, it can be planted as far north as Virginia Beach, Virginia, while in California and Texas it is naturalized as a median strip planting.

Garden History

In his book *Enquiries into Plants* of ca. 300 BC, Theophrastus described, (among plants that affect the mind,) a shrub he called *onotheras*, which modern editors render oleander; "the root of *onotheras* [oleander] administered in wine," he alleges, "makes the temper gentler and more cheerful".

The plant has a leaf like that of the almond, but smaller, and the flower is red like a rose. The plant itself (which loves hilly country) forms a large bush; the root is red and large, and, if this is dried, it gives off a fragrance like wine.

In another mention, of "wild bay" (*Daphne agria*), Theophrastus appears to intend the same shrub.

Willa Cather, in her book *The Song of the Lark*, mentions oleander in the following passage:

This morning Thea saw to her delight that the two oleander trees, one white and one red, had been brought up from their winter quarters in the cellar. There is hardly a German family in the most arid parts of Utah, New Mex-ico, Arizona, but has its oleander trees. However loutish the American--born sons of the family may be, there was never one who refused to give his muscle to the back--breaking task of getting those tubbed trees down into the cellar in the fall and up into the sunlight in the spring. They may strive to avert the day, but they grapple with the tub at last.

Oleander is the official flower of the city of Hiroshima, having been the first to bloom following the atomic bombing of the city in 1945.

Uses and Effects

Ornamental Gardening: Oleander grows well in warm subtropical regions, where it is extensively used as an ornamental plant in landscapes, in parks, and along roadsides. It is drought-tolerant and will tolerate occasional light frost down to –10 °C (14 °F). It is commonly used in landscaping freeway medians in California, Texas and other mild-winter states in the Continental United States because it is upright in habit and easily maintained. Its toxicity renders it deer-resistant. It is tolerant of poor soils and drought. Oleander can also be grown in cooler climates in greenhouses and conservatories, or as indoor plants that can be kept outside in the summer. Oleander flowers are showy and fragrant and are grown for these reasons. Over 400 cultivars have been named, with several additional flower colours not found in wild plants having been selected, including red, purple, pink, and orange; white and a variety of pinks are the most common. Many cultivars also have double flowers. Young plants grow best in spaces where they do not have to compete with other plants for nutrients.

Toxicity: *Nerium oleander* has historically been considered a poisonous plant because some of its compounds may exhibit toxicity, especially to animals, when consumed in high amounts. Among these compounds are oleandrin and oleandrigenin, known as cardiac glycosides, which are known to have a narrow therapeutic index and can be toxic when ingested.

Toxicity studies of animals administered oleander extract concluded that rodents and birds were observed to be relatively insensitive to oleander cardiac glycosides. Other mammals, however, such as dogs and humans, are relatively sensitive to the effects of cardiac glycosides and the clinical manifestations of "glycoside intoxication".

However, despite the common "poisonous" designation of this plant, very few toxic events in humans have been reported. According to the Toxic Exposure Surveillance System (TESS) in 2002 there were 847 human exposures to oleander reported to poison centres in the United States. Despite this exposure level, from 1985 through 2005, only three deaths were reported. One cited death was apparently due to the ingestion of oleander leaves by a diabetic man. His blood indicated a total blood concentration of cardiac glycosides of approximately 20 μg/L which is well above the reported fatal level. Another study reported on the death of a woman who self-administered "an undefined oleander extract" both orally and rectally and her oleandrin tissue levels were 10 to 39 μg/g which were in the high range of reported levels at autopsy. And, finally, one study reported the death of a woman who ingested oleander 'tea'. Few other details were provided. In contrast to consumption of these undefined oleander derived materials, there is no toxicity or deaths reported from topical administration or contact with *Nerium oleander* or specific products derived from them. In reviewing oleander toxicity Lanford and Boor concluded that, except for children who might be at greater risk, "the human mortality associated with oleander ingestion is generally very low, even in cases of moderate intentional consumption (suicide attempts)".

Toxicity studies that have been conducted in dogs and rodents administered oleander extracts by intramuscular (IM) injection indicated that on an equivalent weight basis, doses of an oleander extract with glycosides ten times in excess of those likely to be administered therapeutically to humans are still safe and without any "severe toxicity observed".

Effects of Poisoning

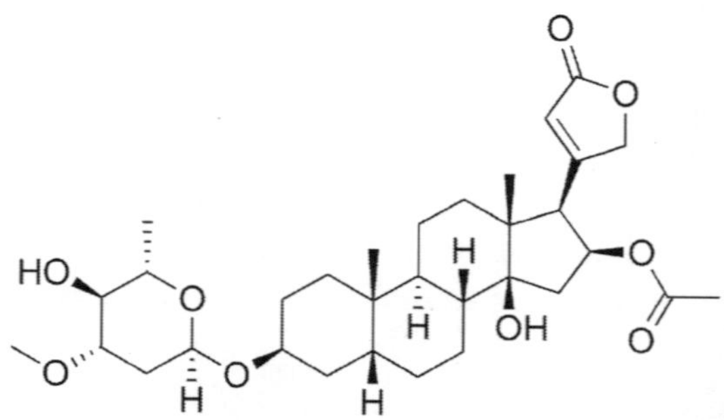

Figure: *Oleandrin, one of the toxins present in Oleander*

Reactions to ingestion of this plant can affect the gastrointestinal system, the heart, and the central nervous system. The gastrointestinal effects can consist of nausea and vomiting, excess salivation, abdominal pain, diarrhea that may or may not contain blood, and especially in horses, colic. Cardiac reactions consist of irregular heart rate, sometimes characterized by a racing heart at first that then slows to below normal further along in the reaction. Extremities may become pale and cold due to poor or irregular circulation. The effect on the central nervous system may show itself in symptoms such as drowsiness, tremors or shaking of the muscles, seizures, collapse, and even coma that can lead to death. Oleander sap can cause skin irritations, severe eye inflammation and irritation, and allergic reactions characterized by dermatitis.

Treatment

Poisoning and reactions to oleander plants are evident quickly, requiring immediate medical care in suspected or known poisonings of both humans and animals. Induced vomiting and gastric lavage are protective measures to reduce absorption of the toxic compounds. Charcoal may also be administered to help absorb any remaining toxins. Further medical attention may be required and will depend on the severity of the poisoning and symptoms. Temporary cardiac pacing will be required in many cases (usually for a few days) till the toxin is excreted.

Digoxin Immune Fab is the best way to cure an oleander poisoning if inducing vomiting has no or minimal success, although it is usually used only for life-threatening conditions due to side-effects.

Drying of plant materials does not eliminate the toxins. It is also hazardous for animals such as sheep, horses, cattle and other grazing animals, with as little as 100 g being enough to kill an adult horse. Plant clippings are especially dangerous to horses, as they are sweet. In July 2009, several horses were poisoned in this manner from the leaves of the plant. Symptoms of a poisoned horse include severe diarrhea and abnormal heartbeat. There is a wide range of toxins and secondary compounds within oleander, and care should be taken around this plant due to its toxic nature. Different names for oleander are used around the world in different locations, so, when encountering a plant with this appearance, regardless of the name used for it, one should exercise great care and caution to avoid ingestion of any part of the plant, including its sap and dried leaves or twigs. The dried or fresh branches should not be used for spearing food, for preparing a cooking fire, or as a food skewer. Many of the oleander relatives, such as the Desert Rose (*Adenium obesum*) found in East Africa, have similar leaves and flowers and are equally toxic.

Therapeutic Efficacy

Drugs derived from *Nerium oleander* have been investigated as a treatment for cancer. According to the American Cancer Society the trials have produced no evidence of benefit, but they did however cause adverse side-effects.

Ecology

Some invertebrates are known to be unaffected by oleander toxins, and feed on the plants. Caterpillars of the Oleander or Polka-Dot Wasp Moth (*Syntomeida epilais*) feed specifically on oleanders and survive by eating only the pulp surrounding the leaf-veins, avoiding the fibres. Larvae of the Common Crow Butterfly (*Euploea core*) also feed on oleanders. The Common Crow larvae retain or modify toxins, making them unpalatable to would-be predators such as birds, but not to other invertebrates such as spiders and wasps.

Wisteria Frutescens

Wisteria frutescens (American Wisteria) is a woody, deciduous, perennial climbing vine of the *Fabaceae* family. It is native to the wet forests and stream banks of the southeastern United States, with a range stretching from the states of Virginia to Texas (Northeast Texas Piney Woods) and extending southeast through Florida, also north to Iowa, Michigan, and New York.

American Wisteria can grow up to 15m long over many supports via powerful clockwise-twining stems. It produces dense clusters of blue-purple, two-lipped, 2-cm-wide flowers on racemes 5–15 cm long in late spring to early summer. These are the smallest racemes produced by any Wisteria. Though it has never been favoured in many gardens for this characteristic, many bonsai artists employ American Wisteria for its manageably-sized flowers, and it is charming as a woodland flowering vine.

The foliage consists of shiny, dark-green, pinnately compound leaves 10–30 cm in length. The leaves bear 9-15 oblong leaflets that are each 2–6 cm long. It also bears numerous poisonous, bean-like seed pods 5–10 cm long that mature in summer and persist until winter; the pods are fuzzy and greenish-tan when young, but shiny brown and smooth when dry. The seeds are large and brown. American Wisteria prefers moist soils. It is considered shade tolerant, but will flower only when exposed to partial or full sun. It grows best in USDA plant hardiness zones 5-9.

Several characteristics distinguish American Wisteria from its Asian counterparts. It grows only two-thirds as tall, its racemes are half as long (the shortest of the Wisteria family), and its bloom time is sometimes shorter than many Asian varieties. Its flowers are not scented, and its seed pods are smooth rather than velvety when mature. Its most redeeming feature is the fact that it is much less invasive than its Asian counterparts, especially the beautiful but ruthless Chinese Wisteria (*Wisteria sinensis*). American Wisteria is very similar to Kentucky Wisteria (*Wisteria macrostachya*) which has been considered a variety of *W. frutescens* but grows somewhat differently and has a fragrance.

Ornamental Trees

Ornamental or specimen trees, both evergreen and deciduous are a creation of beauty in the garden. They are generally planted to form a focal point and provide a sense of shape, texture and colour throughout the year. Decorative and beautiful they are often planted to create a statement and to give the viewer continuous pleasure over many years.

When choosing an ornamental tree, bear in mind the shape, size and style of the garden. Remember that some trees grow very slowly, some grow fast, some trees are small and some are very large. Plant with thought and care.

Delonix Regia

Delonix regia is a species of flowering plant in the family Fabaceae, subfamily Caesalpinioideae. It is noted for its fern-like leaves and flamboyant display of flowers. In many tropical parts of the world it is grown as an ornamental tree and in English it is given the name Royal Poinciana or Flamboyant. It is also one of several trees known as Flame tree.

In India and Pakistan it is known as Gulmohar in Hindi and Urdu. In Persian "gul" means "flower", and "mohr" means "coin" or "stamp". Also "mor" means "peacock", which seems to be most close to physical appearance and beauty of this tree. It is also known there as *Krishnachura* or *Krusnachuda* (Bengali/Oriya: crown of the Krishna) and *Krishnasura* (in Assamese and Bengali). In Vietnam, it is known as Phýãng v) (means "Phoenix's Tail) (Vietnamese), Malinche, and Tabachine. In Guatemala, Antigua Guatemala, it is known as "Llama del Bosque". In Tagalog, it is known as Caballero/Kabalyero.

This species was previously placed in the genus *Poinciana*, named for Phillippe de Longvilliers de Poincy, the 17th century governor of Saint Christophe (Saint Kitts). Because it is a legume, the tree has nitrogen-fixating and soil-improving properties.

Description

The tree's vivid red/vermilion/orange/yellow flowers and bright green foliage make it an exceptionally striking sight.

The Royal Poinciana is found in Madagascar's dry deciduous forests. In the wild it is endangered, but it is widely cultivated elsewhere. In addition to its ornamental value, it is also a useful shade tree in tropical conditions, because it usually grows to a modest height (mostly 5 metres, but it can reach an maximum height of 12 metres) but spreads widely, and its dense foliage provides full shade. In areas with a marked dry season, it sheds its leaves during the drought, but in other areas it is virtually evergreen. Flowers appear in corymbs along and at the ends of branches. Pods are green and flaccid when young and turn dark-brown and woody.

The flowers are large, with four spreading scarlet or orange-red petals up to 8 cm long, and a fifth upright petal called the standard, which is slightly larger and spotted with yellow and white. The naturally occurring variety *flavida* has yellow flowers. Seed pods are dark brown and can be up to 60 cm long and 5 cm wide; the individual seeds, however, are small, weighing around 0.4 g on average. The compound leaves have a feathery appearance and are a characteristic light, bright green. They are doubly pinnate: Each leaf is 30–50 cm long and has 20 to 40 pairs of primary leaflets or pinnae on it, and each of these is further divided into 10-20 pairs of secondary leaflets or pinnules.

Environmental Requirements

The Royal Poinciana requires a tropical or near-tropical climate, but can tolerate drought and salty conditions. The Poinciana prefers an open, free-draining sandy or loamy soil enriched with organic matter. The tree does not like heavy or clay soils and flowers more profusely when kept slightly dry. The Poinciana is very widely grown in the Caribbean, Africa, Northern Australia (the southern extremes previously limited to South East Queensland, although it now grows and blooms successfully in Sydney with flowering trees identified in the suburbs of Petersham, Parramatta, Guildford, Warwick Farm and Kurmond), Hong Kong, the Canary Islands, Mexico, Cyprus, Malta, Thailand, Philippines, Taiwan, southern China. It is the official tree in Vietnam Tainan, Taiwan; Xiamen, Fujian Province, People's Republic of China; and Shantou, Canton Province, People's Republic of China. National Cheng Kung University, a university located in Tainan, put Royal Poinciana on its emblem. It also grows throughout southern Brazil, with ornamental trees in Rio Grande do Sul (Canoas and Porto Alegre).

Geographical Growth Range

Delonix regia is endemic to the western forests of Madagascar, but has been introduced into tropical and sub-tropical regions worldwide. In the continental United States, it grows in South Florida, Southwest Florida, the Rio Grande Valley of South Texas, ranging from the low deserts of Southern Arizona (to as high as Tucson), and Southern California. It also grows in the Dominican Republic, Cuba, Haiti, Hawaii, Mexico (especially in the Yucatan peninsula), Nicaragua, Puerto Rico, U.S. Virgin Islands, Guam and the Commonwealth of the Northern Mariana Islands, where it is the official tree of the islands. It is much loved in the Caribbean; many Dominican & Puerto Rican paintings feature Flamboyant Trees. It can also be found in The Bahamas. The Poinciana is the national flower of St. Kitts and Nevis. The island of Mauritius has widespread distribution of the Royal Poinciana where it announces the coming of the new year.

The Royal Poinciana is regarded as naturalised in many of the locations where it is grown. It is a popular street tree in the suburbs of Brisbane, Australia. The tree is also found in India and Pakistan, where it is referred to as the Gulmohar, or Gul Mohr. In West Bengal (India) and Bangladesh it is called Krishnachura.

The town of Peñuelas, Puerto Rico, located about 12 miles west of Ponce, is nicknamed "*El Valle de los Flamboyanes*" ("The Valley of the Poinciana Trees"), as many Flamboyant trees are found along the surrounding Río Guyanes, Río Macana, and Río Tallaboa Rivers.

In Vietnam, this tree is called "Phýãng vù", or phoenix's tail, and is a popular urban tree in much of Vietnam. Its flowering season is May - July, which coincides with the end of the school year in Vietnam. Because of this timing, the flower of Poinciana is sometimes called the "flower of pupil". Hai Phong city is nicknamed "*Thành phÑ hoa phýãng ðÏ*" ("City of red Poinciana").

Propagation

The Royal Poinciana is most commonly propagated by seeds. Seeds are collected, soaked in warm water for at least 24 hours, and planted in warm, moist soil in a semi-shaded, sheltered position. In lieu of soaking, the seeds can also be 'knicked' or 'pinched' (with a small scissors or nail clipper) and planted immediately. These two methods allow moisture to penetrate the tough outer casing, stimulating germination. The seedlings grow rapidly and can reach 30cm in a few weeks under ideal conditions.

Less common, but just as effective, is propagation by semi-hardwood cuttings. Branches consisting of the current or last season's growth can be cut into 30cm sections and planted in a moist potting mixture. This method is slower than seed propagation (cuttings take a few months to root) but is the preferred method for ensuring new trees are true to form. As such, cuttings are a particularly common method of propagation for the rarer yellow-flowering variety of the tree.

Flowering Season

- Bangladesh: April-May
- South Florida: May–June
- Egypt: May-June
- Vietnam: May–July
- Caribbean: May–September
- Indian subcontinent (India, Pakistan, Bangladesh): April–June
- Australia: December–February
- Northern Mariana Islands: March–June

- United Arab Emirates: May-July
- Brazil: November–February
- Southern Sudan: March–May
- Thailand: April–May
- Philippines: April-May
- Zambia and Zimbabwe: October–December
- Hong Kong: May-June
- Mauritius: November-December

Cherry Blossom

A cherry blossom is the flower of any of several trees of genus *Prunus*, particularly the Japanese Cherry, *Prunus serrulata*, which is sometimes called sakura after the Japanese.

Cherry blossom is speculated to be native to the Himalayas. Currently it is widely distributed, especially in the temperate zone of the Northern Hemisphere such as: Europe, West Siberia, China, Japan, United States, etc.

Many of the varieties that have been cultivated for ornamental use do not produce fruit. Edible cherries generally come from cultivars of the related species *Prunus avium* and *Prunus cerasus*.

Flower Viewing

"Hanami" is the centuries-old practice of picnicking under a blooming *sakura* or *ume* tree. The custom is said to have started during the Nara Period (710–794) when it was *ume* blossoms that people admired in the beginning. But by the Heian Period (794–1185), cherry blossoms came to attract more attention and *hanami* was synonymous with *sakura*. From then on, in both waka and haiku, "flowers" (±, *hana*) meant "cherry blossoms". The custom was originally limited to the elite of the Imperial Court, but soon spread to samurai society and, by the Edo period, to the common people as well. Tokugawa Yoshimune planted areas of cherry blossom trees to encourage this. Under the *sakura* trees, people had lunch and drank sake in cheerful feasts.

Every year the Japanese Meteorological Agency and the public track the *sakura zensen* (cherry blossom front) as it moves northward up the archipelago with the approach of warmer weather via nightly forecasts following the weather segment of news programmes. The blossoming begins in Okinawa in January and typically reaches Kyoto and Tokyo at the end of March or the beginning of April. It proceeds into areas at the higher altitudes and northward, arriving in Hokkaidô a few weeks later. Japanese pay close attention to these forecasts and turn out in large numbers at parks, shrines, and temples with family and friends to hold flower-viewing parties. *Hanami* festivals celebrate the beauty of the cherry blossom and for many are a chance to relax and enjoy the beautiful view. The custom of *hanami* dates back many centuries in Japan. The eighth-century chronicle *Nihon Shoki* records *hanami* festivals being held as early as the third century CE.

Most Japanese schools and public buildings have cherry blossom trees outside of them. Since the fiscal and school year both begin in April, in many parts of Honshû, the first day of work or school coincides with the cherry blossom season.

The Japan Cherry Blossom Association developed a list of Japan's Top 100 Cherry Blossom Spots with at least one location in every prefecture.

Symbolism

In Japan, cherry blossoms also symbolize clouds due to their nature of blooming *en masse*, besides being an enduring metaphor for the ephemeral nature of life, an aspect of Japanese cultural tradition that is often associated with Buddhistic influence, and which is embodied in the concept of *mono no aware*. The association of the cherry blossom with *mono no aware* dates back to 18th-century scholar Motoori Norinaga. The transience of the blossoms, the extreme beauty and quick death, has often been associated with mortality; for this reason, cherry blossoms are richly symbolic, and have been utilized often in Japanese art, manga, anime, and film, as well as at musical performances for ambient effect. There is at least one popular folk song, originally meant for the shakuhachi (bamboo flute), titled "Sakura", and several pop songs. The flower is also represented on all manner of consumer goods in Japan, including kimono, stationery, and dishware.

The *Sakurakai* or Cherry Blossom Society was the name chosen by young officers within the Imperial Japanese Army in September 1930 for their secret society established with the goal of reorganizing the state along totalitarian militaristic lines, via a military coup d'état if necessary.

During World War II, the cherry blossom was used to motivate the Japanese people, to stoke nationalism and militarism among the populace. Even prior to the war, they were used in propaganda to inspire "Japanese spirit," as in the "Song of Young Japan," exulting in "warriors" who were "ready like the myriad cherry blossoms to scatter." In 1932, Akiko Yosano's poetry urged Japanese soldiers to endure sufferings in China and compared the dead soldiers to cherry blossoms. Arguments that the plans for the Battle of Leyte Gulf, involving all Japanese ships, would expose Japan to serious danger if they failed, were countered with the plea that the Navy be permitted to "bloom as flowers of death." The last message of the forces on Peleliu was "Sakura, Sakura" — cherry blossoms. Japanese pilots would paint them on the sides of their planes before embarking on a suicide mission, or even take branches of the trees with them on their missions. A cherry blossom painted on the side of the bomber symbolized the intensity and ephemerality of life; in this way, the aesthetic association was altered such that falling cherry petals came to represent the sacrifice of youth in suicide missions to honour the emperor. The first kamikaze unit had a subunit called *Yamazakura* or wild cherry blossom. The government even encouraged the people to believe that the souls of downed warriors were reincarnated in the blossoms.

In its colonial enterprises, imperial Japan often planted cherry trees as a means of "claiming occupied territory as Japanese space".

Cherry blossoms are a prevalent symbol in Irezumi, the traditional art of Japanese tattoos. In tattoo art, cherry blossoms are often combined with other classic Japanese symbols like koi fish, dragons or tigers.

Japan has a wide variety of cherry blossoms (sakura); well over 200 cultivars can be found there. The most popular variety of cherry blossom in Japan is the *Somei Yoshino*. Its flowers are nearly pure white, tinged with the palest pink, especially near the stem. They bloom and usually fall within a week, before the leaves come out. Therefore, the trees look nearly white from top to bottom. The variety takes its name from the village of Somei (now part of Toshima in Tokyo). It was developed in the mid- to late-19th century at the end of the Edo period and the beginning of the Meiji period. The *Somei Yoshino* is so widely associated with cherry blossoms that *jidaigeki* and other works of fiction often depict the variety in the Edo period or earlier; such depictions are anachronisms.

Winter sakura or *fuyuzakura* (*Prunus subhirtella autumnalis*) begins to bloom in the fall and continues blooming sporadically throughout the winter. It is said to be a cross between *edohiganzakura*, the Tokyo Higan cherry (*P. incisa*) and *mamezakura* (*P. pendula*).

Other categories include *yamazakura*, *yaezakura*, and *shidarezakura*. The *yaezakura* have large flowers, thick with rich pink petals. The *shidarezakura*, or weeping cherry, has branches that fall like those of a weeping willow, bearing cascades of pink flowers.

By Country

Australia

During World War II, a prisoner of war (POW) camp near the town of Cowra in New South Wales, Australia was the site of one of the largest prison escapes of the war, on 5 August 1944. During the Cowra breakout and subsequent rounding up of POWs, four Australian soldiers and 231 Japanese soldiers died and 108 prisoners were wounded. The Japanese War Cemetery holding the dead from the Breakout was tended to after WWII by members of the Cowra RSL and ceded to Japan in 1963. In 1971 the Cowra Tourism Development decided to celebrate this link to Japan, and proposed a Japanese Garden for the town. The Japanese government agreed to support this development as a sign of thanks for the respectful treatment of their war dead; the development also received funding from the Australian government and private entities.

The garden was designed by Ken Nakajima (1914–2000), a world-renowned designer of Japanese gardens at the time. The first stage was opened in 1979, with a second stage opened in 1986.

The gardens were designed in the style of the Edo period and are a *kaiyû-shiki* or strolling garden. They are designed to show all of the landscape types of Japan. At five hectares (12 acres), the Cowra Japanese Garden is the largest Japanese garden in the Southern Hemisphere. An annual cherry blossom festival is a major event in Cowra's tourism calendar and is held in the gardens during September.

Brazil

With the Japanese diaspora to Brazil, many immigrants brought seedlings of cherry trees. In São Paulo State, home to the largest Japanese community outside Japan, it is common to find them in Japan-related facilities and in some homes, usually of the cultivars *Prunus serrulata* 'Yukiwari' and *Prunus serrulata* var. *lannesiana* 'Himalaya'. In the Parana State (in southern Brazil), many cities received many of these immigrants, who planted the trees, as in Apucarana, Maringá, Cascavel and especially in the capital city of Curitiba.

In the capital city of Parana, the first seedlings were brought by Japanese immigrants in the first half of the 20th century, but large quantities of them were only planted from the 1990s, with the opening of the Botanical Garden of Curitiba. Nowadays the seedlings are produced locally and used in afforestation of streets and squares – as in the Japanese Square, where there are more than 30 cherry trees around the square which were sent by the Japanese Empire to Curitiba.

Canada

Vancouver, BC is famous for its thousands of cherry trees (estimated 50,000) lining many streets and in many parks, including Queen Elizabeth Park and Stanley Park. Vancouver holds the Vancouver Cherry Blossom Festival every year. With multiple varieties and a temperate climate, they begin to bloom in February yearly and peak in April.

High Park in Toronto, Ontario features many Somei-Yoshino cherry trees (the earliest species to bloom and much loved by the Japanese for their fluffy white flowers) that were given to Toronto by Japan in 1959. Through the Sakura Project, the Japanese Consulate donated a further 34 cherry trees to High Park in 2001, plus cherry trees to various other locations like Exhibition Place, McMaster University, York University (near Calumet College and on Ottawa Road near McLaughlin College) and the University of Toronto's main (next to Robarts Library) and Scarborough campuses. Niagara Falls also has many near the Falls itself. Royal Botanical Gardens in Burlington and Hamilton was also the recipient of a number of Somei-Yoshino cherry trees that were donated by the Consulate-General of Japan in Toronto as part of the Sakura Project. The trees are located in the Arboretum and the Rock Garden and were planted to celebrate the continual strengthening of friendship between Japan and Canada. Peak bloom time at Royal Botanical Gardens is normally around the last week of April or the first week of May.

China

Cherry trees naturally grow in the middle northern or southern part of China, the area nearby the sea. However, the most famous cherry blossom parks in China reflect Japan's brief occupation of parts of China during the first half of the 20th century or the donation from Japan thereafter:

- Longwangtang Cherry Blossom Park in Lushun, Dalian, Liaoning
- East Lake Cherry Blossom Park near Wuhan University, in Donghu District, Wuhan, Hubei
- Wuhan University, in Donghu District, Wuhan, Hubei
- Nanshan Botanical Garden in Nan'an District, Chongqing

Germany

The cherry blossom is a major tourist attraction in Germany's Altes Land orchard region. The largest Hanami in Germany, in Hamburg, with Japanese-style fireworks, organized by the German-Japanese society, draws tens of thousands spectators every spring.

India

In India, cherry blossom is attraction as well, most notably in eastern and northeastern India and in various British-era botanical gardens especially in cooler climatic zones such as in the Himalayas, Nilgiri Hills, Garo Hills, Khasi Hills as well as in some hill stations in the Western Ghats and in the northern districts of West Bengal of Jalpaiguri and Darjeeling.

The Netherlands

In the year 2000, the Japan Women's Club (JWC) donated 400 cherry blossom trees to the city of Amstelveen. The trees have been planted in the Cherry blossom park in the Amsterdamse Bos. A special detail is that every tree has a name — 200 trees have female Japanese names, 200 trees have female Dutch names.

South Korea

Watching of cherry blossom was introduced to Korea during Japanese rule. The festivals continued even after the Japanese surrendered in WWII, but have been contentious, and many cherry trees have been destroyed because they were seen as symbols of the occupation. However, there has been considerable confusion about the origin of the cultivated Japanese cherry trees and the differences between them and native Korean trees. Certain trees at Seoul's Gyeongbok Palace were cut down to celebrate the fiftieth anniversary of Japanese surrender in World War II. Although Cherry blossoms are already indigenous to Korea, Japan had planted trees on sacred and provocative locations in the Palace. Once the offending trees were cut down the festival continued with the indigenous trees. The cherry blossom festival at Gyeongbok Palace is one of a number of such festivals across Korea and is prominently advertised to tourists.

Turkey

In 2005, Japanese cherry trees were presented by Japan to the Nezahat Gokyigit Botanical Garden in Istanbul, Turkey. Each tree represents one sailor of the frigate *Ertogrul* which was a famous frigate of the Ottoman Turkish navy. On the way back from a goodwill visit to Japan in 1890 she ran into a typhoon and sank with loss of 587 Ottoman Turkish sailors. That loss is being remembered at every anniversary. The Japanese cherry trees represent memory of those who died and provide remembrance.

United Kingdom

Batsford Arboretum in Gloucestershire (England), holds the national collection of Japanese village cherries, sato-sakura group. Keele

University in Staffordshire (England), has one of the UK's largest collections of flowering cherries, with more than 150 varieties.

United States

Branch Brook Park, which is shared by the City of Newark and the Township of Belleville New Jersey, is the oldest county park in the United States and is home to the nation's' largest collection of cherry blossom trees, with about 4,300.

Japan gave 3,020 cherry blossom trees as a gift to the United States in 1912 to celebrate the nations' then-growing friendship, replacing an earlier gift of 2000 trees which had to be destroyed due to disease in 1910. These trees were planted in Sakura Park in Manhattan and line the shore of the Tidal Basin in Washington, D.C. The first two original trees were planted by first lady Helen Taft and Viscountess Chinda on the bank of the Tidal Basin. The gift was renewed with another 3,800 trees in 1965. In Washington, D.C. the cherry blossom trees continue to be a popular tourist attraction (and the subject of the annual National Cherry Blossom Festival) when they reach full bloom in early spring.

Also, Balboa Park of San Diego has 2,000 cherry blossom trees that blossom in mid to late March. In Los Angeles, over 2,000 trees are located at Lake Balboa in Van Nuys. These trees were donated by an anonymous Japanese benefactor and were planted in 1992. They originated from a single parent tree and were developed to grow in warm climates.

Philadelphia is also home to over 2000 flowering Japanese cherry trees, half of which were a gift from the Japanese government in 1926 in honour of the 150th anniversary of American independence, with the other half planted by the Japan America Society of Greater Philadelphia between 1998 and 2007. Philadelphia's cherry blossoms are located within Fairmount Park, and the annual Subaru Cherry Blossom Festival of Greater Philadelphia celebrates the blooming trees. The University of Washington in Seattle also has cherry blossoms in its Quad.

Other US cities have an annual Cherry Blossom Festival (or Sakura Matsuri), including the International Cherry Blossom Festival in Macon, Georgia, which features over 300,000 cherry trees. The Brooklyn Botanic Garden in New York City also has a large, well-attended festival. Portsmouth, New Hampshire, is the site of the peace conference that produced the Treaty of Portsmouth, for which the original Washington, DC cherry trees were given in thanks. Several cherry trees planted on the bank of the tidal pond next to Portsmouth City Hall were the gift of Portsmouth's Japanese sister city of Nichinan—the hometown of Marquis Komura Jutarô, Japan's representative at the conference.

Ohio University in Athens, Ohio, has 200 somei yoshino trees, a gift from its sister institution, Japan's Chubu University.

Culinary Use

Cherry blossoms and leaves are edible and both are used as food ingredients in Japan:

- The blossoms are pickled in salt and umezu (ume vinegar), and used for coaxing out flavour in wagashi, (a traditional Japanese confectionery,) or anpan, (a Japanese sweet bun, most-commonly filled with red bean paste.)
- Salt-pickled blossoms in hot water are called sakurayu, and drunk at festive events like weddings in place of green tea.
- The leaves, mostly from the Ôshima cherry because of the softness, are also pickled in salted water and used for sakuramochi.

Since the leaves contain coumarin, which is toxic in large doses, it is not recommended to eat them in great quantities.

Palms

The Arecaceae are a botanical family of perennial lianas, shrubs, and trees commonly known as palm trees. (Due to historical usage, the family is alternatively called Palmae or Palmaceae.) They are flowering plants, the only family in the monocot order Arecales. Roughly 202 genera with around 2600 species are currently known, most of them restricted to tropical, subtropical, and warm temperate climates. Most palms are distinguished by their large, compound, evergreen leaves arranged at the top of an unbranched stem. However, many palms are exceptions, and in fact exhibit an enormous diversity in physical characteristics. As well as being morphologically diverse, palms also inhabit nearly every type of habitat within their range, from rainforests to deserts. Palms are among the best known and most extensively cultivated plant families. They have been important to humans throughout much of history. Many common products and foods are derived from palms, and palms are also widely used in landscaping for their exotic appearance, making them one of the most economically important plants. In many historical cultures, palms were symbols for such ideas as victory, peace, and fertility. Today, palms remain a popular symbol for the tropics and vacations.

Morphology

Whether as shrubs, trees, or vines, palms have two methods of growth: solitary or clustered. The common representation is that of a solitary shoot ending in a crown of leaves. This monopodial character may be exhibited by prostrate, trunkless, and trunk-forming members. Some common palms restricted to solitary growth include *Washingtonia* and *Roystonea*. Palms may instead grow in sparse though dense clusters. The trunk develops an axillary bud at a leaf node, usually near the base, from which a new shoot emerges. The new shoot, in turn, produces an axillary bud and a clustering habit results. Exclusively sympodial genera include many of the rattans, *Guihaia*, and *Rhapis*. Several palm genera have both solitary and clustering members. Palms which are usually solitary may grow in clusters, and *vice versa*. These aberrations suggest the habit operates on a single gene.

Palms have large, evergreen leaves that are either palmately ('fan-leaved') or pinnately ('feather-leaved') compound and spirally arranged at the top of the stem. The leaves have a tubular sheath at the base that usually splits open on one side at maturity. The inflorescence is a spadix or spike surrounded by one or more bracts or spathes that become woody at maturity. The flowers are generally small and white, radially symmetric, and can be either uni- or bisexual. The sepals and petals usually number three each, and may be distinct or joined at the base. The stamens generally number six, with filaments that may be separate, attached to each other, or attached to the pistil at the base. The fruit is usually a single-seeded drupe (sometimes berry-like) but some genera (e.g. *Salacca*) may contain two or more seeds in each fruit.

The Arecaceae are notable among monocots for their height and for the size of their seeds, leaves, and inflorescences. *Ceroxylon quindiuense*, Colombia's national tree, is the tallest monocot in the world, reaching up to 60 metres tall. The *coco de mer* (*Lodoicea maldivica*) has the largest seeds of any plant, 40–50 cm in diametre and weighing 15–30 kg each. Raffia palms (*Raphia* spp.) have the largest leaves of any plant, up to 25 m long and 3 m wide. The *Corypha* species have the largest inflorescence of any plant, up to 7.5 m tall and containing millions of small flowers. *Calamus* stems can reach 200 m in length.

Range and Habitat

Most palms grow in the tropics. They are abundant throughout the tropics, and thrive in almost every habitat therein. Their diversity is highest in wet, lowland tropical forests, especially in ecological "hotspots" such as Madagascar, which has more endemic palms than all of Africa. Colombia may have the highest number of palm species in one country. Palms are most commonly seen throughout Africa, South America, Mexico, and areas of the United States such as North Carolina, Texas,

Florida, Arizona, New Mexico, California, Georgia, Louisiana, Alabama, Mississippi, South Carolina, Puerto Rico, and Hawaii.

Only an estimated 130 palm species grow naturally beyond the tropics, mostly in the subtropics. The northernmost native palm is *Chamaerops humilis*, which reaches 44°N latitude in southern France. The southernmost palm is the *Rhopalostylis sapida*, which reaches 44°S on the Chatham Islands where an oceanic climate prevails. Some palms, such as the *Trachycarpus fortunei*, grow well under cultivation as far north as over 50°N in oceanic climates (Ireland, Scotland and the Pacific Northwest, from Oregon to Vancouver). They are located in other countries, such as the United Kingdom, most notably in coastal towns such as Paignton, Margate, Blackpool, Ellesmere Port, and Hull.

Palms inhabit a variety of ecosystems. More than two-thirds of palm species live in tropical forests, where some species grow tall enough to form part of the canopy and shorter ones form part of the understory. Some species form pure stands in areas with poor drainage or regular flooding, including *Raphia hookeri* which is common in coastal freshwater swamps in West Africa. Other palms live in tropical mountain habitats above 1000 m, such as those in the genus *Ceroxylon* native to the Andes. Palms may also live in grasslands and scrublands, usually associated with a water source, and in desert oases such as the date palm. A few palms are adapted to extremely basic lime soils, while others are similarly adapted to extreme potassium deficiency and toxicity of heavy metals in serpentine soils.

Taxonomy

Palms are a monophyletic group of plants, meaning the group consists of a common ancestor and all its descendants. Extensive taxonomic research on palms began with botanist H.E. Moore, who organized palms into 15 major groups based mostly on general morphological characteristics. The following classification, proposed by N.W. Uhl and J. Dransfield in 1987, is a revision of Moore's classification that organizes palms into six subfamilies. A few general traits of each subfamily are listed.

The Coryphoideae are the most diverse subfamily, and are a paraphyletic group, meaning all members of the group share a common ancestor, but the group does not include all the ancestor's descendants. Most palms in this subfamily have palmately lobed leaves and solitary flowers with three, or sometimes four carpels. The fruit normally develops from only one carpel. Subfamily Calamoideae includes the climbing palms, such as rattans. The leaves are usually pinnate; derived characters (synapomorphies) include spines on various organs, organs specialized for climbing, an extension of the main stem of the leaf-bearing reflexed spines, and overlapping scales covering the fruit and ovary. Subfamily Nypoideae contains only one species, *Nypa fruticans*, which has large, pinnate leaves. The fruit is unusual in that it floats, and the stem is dichotomously branched, also unusual in palms. Subfamily Ceroxyloideae has small to medium-sized flowers, spirally arranged, with a gynoecium of three joined carpels. The Arecoideae are the largest subfamily, with six diverse tribes (Areceae - Caryoteae - Cocoeae - Geonomeae - Iriarteeae - Podococceae) containing over 100 genera. All tribes have pinnate or bipinnate leaves and flowers arranged in groups of three, with a central pistillate and two staminate flowers. The Phytelephantoideae are a monoecious subfamily. Members of this group have distinct monopodial flower clusters. Other distinct features include a gynoecium with five to 10 joined carpels, and flowers with more than three parts per whorl. Fruits are multiple-seeded and have multiple parts.

Figure: *Walsh River palms*

Currently, few extensive phylogenetic studies of Arecaceae exist. In 1997, Baker *et al.* explored subfamily and tribe relationships using chloroplast DNA from 60 genera from all subfamilies and tribes. The results strongly showed the Calamoideae are monophyletic, and Ceroxyloideae and Coryphoideae are paraphyletic. The relationships of

Arecoideae are uncertain, but they are possibly related to Ceroxyloideae and Phytelephantoideae. Studies have suggested the lack of a fully resolved hypothesis for the relationships within the family is due to a variety of factors, including difficulties in selecting appropriate outgroups, homoplasy in morphological character states, slow rates of molecular evolution important for the use of standard DNA markers, and character polarization. However, hybridization has been observed among *Orbignya* and *Phoenix* species, and using chloroplast DNA in cladistic studies may produce inaccurate results due to maternal inheritance of the chloroplast DNA. Chemical and molecular data from non-organelle DNA, for example, could be more effective for studying palm phylogeny.

Evolution

The Arecaceae are the first modern family of monocots clearly represented in the fossil record. Palms first appear in the fossil record around 80 million years ago, during the late Cretaceous period. The first modern species, such as *Nypa fruticans* and *Acrocomia aculeata*, appeared 94 million years ago, confirmed by fossil *Nypa* pollen dated to 94 million years ago. Palms appear to have undergone an early period of adaptive radiation. By 60 million years ago, many of the modern, specialized genera of palms appeared and became widespread and common, much more widespread than their range today. Because palms separated from the monocots earlier than other families, they developed more intrafamilial specialization and diversity. By tracing back these diverse characteristics of palms to the basic structures of monocots, palms may be valuable in studying monocot evolution. Several species of palms have been identified from flowers preserved in amber, including *Palaeoraphe dominicana* and *Roystonea palaea*.

Evidence can also be found in samples of petrified palmwood.

Uses

Human use of palms is as old or older than human civilization itself, starting with the cultivation of the date palm by Mesopotamians and other Middle Eastern peoples 5000 years or more ago. Date wood, pits for storing dates, and other remains of the date palm have been found in Mesopotamian sites. The date palm had a tremendous effect on the history of the Middle East. W.H. Barreveld wrote:

> *"One could go as far as to say that, had the date palm not existed, the expansion of the human race into the hot and barren parts of the "old" world would have been much more restricted. The date palm not only provided a*

concentrated energy food, which could be easily stored and carried along on long journeys across the deserts, it also created a more amenable habitat for the people to live in by providing shade and protection from the desert winds (Fig. 1). In addition, the date palm also yielded a variety of products for use in agricultural production and for domestic utensils, and practically all parts of the palm had a useful purpose."

An indication of the importance of palms in ancient times is that they are mentioned more than 30 times in the Bible, and at least 22 times in the Quran. Arecaceae have great economic importance, including coconut products, oils, dates, palm syrup, ivory nuts, carnauba wax, rattan cane, raffia and palm wood. Along with dates mentioned above, members of the palm family with human uses are numerous.

Some Ornamental Grasses

Tussock (grass)

Tussock grasses or bunch grasses are found as native plants in natural ecosystems, as forage in pastures, and as ornamental grasses in gardens. Tussock and bunch grasses, in the Poaceae family, are grasses that usually grow as singular plants in clumps, tufts, hummocks, or bunches, rather than forming a sod or lawn, in meadows, grasslands, and prairies. As perennial plants usually, they live more than one season.

Many species have long roots that may reach 2-metre (6.6 ft) or more into the soil, which can aid slope stabilization, erosion control, and soil porosity for precipitation absorption. Also, their roots can reach moisture more deeply than other grasses and annual plants during seasonal or climatic droughts. The plants provide habitat and food for insects (including Lepidoptera), birds, small animals and larger herbivores, and support beneficial soil mycorrhiza. The leaves supply material, such as for basket weaving, for indigenous peoples and contemporary artists.

Tussock and bunch grasses occur in almost any habitat where other grasses are found, including: grasslands, savannas and prairies, wetlands and estuaries, riparian zones, shrublands and scrublands, woodlands and forests, montane and alpine zones, tundra and dunes, and deserts.

Fire Resistance

In western North American wildfires, bunch grasses tend to smolder and not ignite into flames, unlike invasive species of annual grasses that contribute to a fire's spreading.

Bermuda Grass

Bermudagrass (*Cynodon* spp.) is a major turf species for sports fields, lawns, parks, golf courses, and general utility turfs in Australia, Africa, India, South America and the Southern region of the United States. It is found in over 100 counties throughout the tropical and subtropical areas of the world.

The genus *Cynodon* comprises nine species with *C. dactylon* being the most widespread. That C. dactylon is a tetraploid with broad genetic variability serves to explain its widespread distribution. Other *Cynodon* species have a more limited natural distribution and are often restricted to one particular habitat. *C. dactylon* is highly fertile, whereas the diploid species such as *C. transvaalensis* rarely produce viable seed.

C. dactylon (L.) Pers is commonly called bermudagrass in many areas of the world. But, it is also known by numerous other names including "Kweekgras" (S. Africa), couch grass (Australia and Africa), devil's grass (India) and gramillia (Argentina). The variety of names given this species attests to its wide distribution and to the fact that it is the object of abuse and scorn.

In addition to being a widely used species for forage and turf, *C. dactylon* is a serious weed in many crops. Being a vigorous, stoloniferous grass, it rapidly invades crops in high rainfall or irrigated areas. *C. dactylon* is ranked among the three most troublesome weeds in sugarcane, cotton, corn and vineyards in many countries. It is a difficult weed to eradicate because of its seed production and deep rhizomes.

In the United States the distribution of bermudagrass extends from New Jersey and Maryland southward to Florida and westward to Kansas and Texas. Under irrigation its distribution extends westward to southern New Mexico, Arizona and to most major valleys in California. The development of more cold tolerant turf-type varieties of bermudagrass such as U-3 and Midiron has increased interest in the species near its northern limits. Low winter temperature is the factor that limits the northward distribution of bermudagrass.

Description. Bermudagrass (*C. dactylon* (L.) Pers.) is a highly variable, sod forming perennial that spreads by stolons, rhizomes and seed. Stolons of bermudagrass readily root at the nodes. Lateral buds develop at the nodes to produce erect or ascending stems that reach 5 to 40 cm (rarely over 90 cm) in height. In most *Cynodon* sp., leaves are borne on stems with long internodes alternating with one or more very short internodes. This characteristic gives the impression that the species has multiple-leaved nodes. Leaf sheaths are compressed to round, loose, split, smooth, sparsely hairy, up to 15 cm long, and with a tuft of hairs 2 to 5 mm long. Auricles are absent. Collar is continuous, narrow, glabrous and hairy on margins. Leaf blades are 2 to 16 cm long, 1.5 to 5 mm wide, smooth to sparsely pubescent, folded or loosely rolled in the bud and sharply-pointed. The inflorescence consists of 3 to 7 spikes in a single whorl in a fingerlike arrangement and 3 to 10 cm long. In robust forms there may be up to 10 spikes, sometimes in two whorls. Spikelets are 2 to 3 mm long, in 2 rows tightly appressed to one side of the rachis; glumes are to ° the length of spikelet; lemma is boot-shaped, acute with fringe of hairs on the keel and longer than the glume; seed is 1.5 mm long, oval, straw to red-coloured and free within the lemma and palea.

Bermudagrass has a fibrous, perennial root system with vigorous, deep rhizomes. Roots are produced at the nodes after new leaves or tillers are produced during the growing season and after new shoots are produced in the spring. Mature roots are yellow to brown while new roots are white. Mature roots deteriorate throughout the growing season and new roots are produced continuously. Root production and dieback has been reported to be particularly high in the spring at the onset of shoot production.

Adaptation and Use. Bermudagrass is a warm season perennial species adapted to tropical and subtropical climates. It grows best under extended periods of high temperatures, mild winters and moderate to high rainfall. Temperature is the main environmental factor that limits its adaptability to tropical and subtropical areas of the world. The northern limits of bermudagrass extend into the transitional zone of the United States where low temperatures seldom drop below 10°F. In general, temperatures below 30°F kill the leaves and stems of bermudagrass. Research has demonstrated that bermudagrass will continue to grow with night temperatures as low as 34°F if day temperatures are near 70°F. However, when average temperatures drop below 50°F growth stops and the grass begins to discolour. At the onset of low temperatures in the fall and winter, bermudagrass begins to discolour, protein fractions change in composition and reserve carbohydrates increase in the stems and rhizomes. After the first killing frost, leaves and stems of bermudagrass remain dormant until average daily temperatures rise above 50°F for several days. The roots and rhizomes of bermudagrass continue to grow several weeks after the leaves and stems stop growth.

In warm frost-free climates bermudagrass remains green throughout the year, but growth is significantly reduced at the onset of cool nights.

The species makes the best growth where average daily temperatures are above 75°F. Optimum daytime temperature for bermudagrass is between 95° and 100°F.

Soil temperature, as influenced by air temperature, is also important to the growth and development of bermudagrass turf. Soil temperatures above 65°F are required for significant growth of rhizomes, roots and stolons. Optimum soil temperature for root growth is around 80°F.

Bermudagrass has a high light requirement and does not grow well under low light (shaded) conditions. The duration of the light period (day length) also influences growth and development of bermudagrass. Both increased light intensity and day length increase rhizome, stolon and leaf growth in bermudagrass. At low light intensities (less than 60% full sunlight) bermudagrass develops narrow, elongated leaves; thin upright stems; elongated internodes and weak rhizomes. Consequently, bermudagrass develops a very sparse turf under moderately shaded conditions.

Bermudagrass is found in tropical and subtropical climates with 25 to 100 inches of annual rainfall, but it also survives in arid climates along waterways and in irrigated areas. Where annual rainfall is below 20 inches per year, bermudagrass requires irrigation to survive. Bermudagrass develops into a semidormant state during very dry conditions, but has the capability of surviving extreme droughts. Rhizomes of bermudagrass can lose 50% or more of their weight and still recover when favourable moisture develops. Generally, common bermudagrass, or tetraploids of *C. dactylon*, have the deepest root and rhizome penetration and better withstand prolonged drought periods.

Common bermudagrass also has the characteristic of producing seedheads under stress conditions such as drought. Thus, the seeds provide another method by which the species can survive extreme drought. Some natural biotypes of *C. dactylon* produce numerous seeds. The seeds are very small with about 2,000,000 seeds per pound.

Bermudagrass grows well on a wide variety of soils from heavy clays to deep sands, provided fertility is not limiting. It tolerates both acid and alkaline soil conditions and is highly tolerant to saline conditions. Bermudagrass survives some flooding but does best on well-drained sites. Although it may persist under low fertility, bermudagrass has a high nitrogen requirement for good quality turf.

Bermudagrass has numerous turf uses. It has been suggested that if ever a plant deserved a monument for its service to mankind, it was bermudagrass for what it has done to prevent soil erosion, to stabilize ditch banks, roadsides and airfields, to beautify landscapes and to provide a smooth, resilient playing surface for sports fields and playgrounds. Bermudagrass also provides hay and pasture for livestock throughout the tropical and subtropical areas of the world.

Epipremnum Aureum Ivy

Epipremnum aureum is a species of flowering plant in the family Araceae, with a broad native Old World distribution. Native range extends from Northern Australia through Malaysia and Indochina into China, Japan and India. The species has become naturalised in tropical and sub-tropical forests worldwide, where it has caused severe ecological damage in some cases. The plant has a multitude of common names including Australian native monstera, centipede tongavine, devil's ivy, golden pothos, hunter's robe, ivy arum, money plant, silver vine, Solomon Islands ivy and taro vine. It is sometimes mistakenly labelled as a Philodendron in plant stores.

E. aureum is an evergreen vine growing to 20 m (66 ft) tall, with stems up to 4 cm (2 in) in diametre, climbing by means of aerial roots which adhere to surfaces. The leaves are alternate, heart-shaped, entire on juvenile plants, but irregularly pinnatifid on mature plants, up to 100 cm (39 in) long and 45 cm (18 in) broad (juvenile leaves much smaller, typically under 20 cm (8 in) long). The flowers are produced in a spathe up to 23 cm (9 in) long. This plant produces trailing stems when it climbs up trees and these take root when they reach the ground and grow along it. The leaves on these trailing stems grow up to 10 cm (4 in) long and are the ones normally seen on this plant when it is cultivated as a pot plant.

Cultivation and Uses

In temperate regions it is a popular houseplant with numerous cultivars selected for leaves with white, yellow, or light green variegation. It is often used in decorative displays in shopping centres, offices, and other public locations largely because it requires little care and is also attractively leafy. It is also efficient at removing indoor pollutants such as formaldehyde, xylene, and benzene. A study found that this effect lessened the higher the molecular weight of the polluting substance. As a houseplant it can reach a height of 2 m (6 ft 7 in) or more, given suitable support. It has gained the Royal Horticultural Society's Award of Garden Merit.

The plant is sometimes used in aquariums, placed on top of the aquarium and allowed to grow roots in the water. This is beneficial to the plant and the aquarium as it absorbs many nitrates and uses them for growth.

Toxicity

The plant is listed as "toxic to cats, toxic to dogs" by the ASPCA, because of the presence of insoluble raphides. Care should be taken to ensure the plant is not consumed by house pets or children. Symptoms may include oral irritation, vomiting, and difficulty swallowing.

Invasive Species

E. aureum can become a highly invasive species when introduced into tropical countries where it is not native. In Sri Lanka it overgrows several hectares of the Udawatta Kele Sanctuary in Kandy. Having no natural enemies, it completely overgrows the forest floor as well as the trunks of trees, causing severe ecological disruption.

It also invades the Kurulukele Forest Reserve in Kegalla and other places where it has been planted as a decorative plant, or to hold steep banks along roads. It is on the USDA list for Florida Exotic Pest Control Council list 1999 as an invasive species.

Ophiopogon Jaburana Varegated

Ophiopogon (lilyturf) is a genus of evergreen perennial plants with about 65 species, native to warm temperate to tropical east, southeast, and south Asia. Despite their grasslike appearance, they are not closely related to the true grasses Poaceae. The name of the genus is derived from Greek *ophis*, "snake", and *pogon*, "beard", most probably referring to its leaves and tufted growth. In the APG III classification system, it is placed in the family Asparagaceae, subfamily Nolinoideae (formerly the family Ruscaceae). Like many lilioid monocots, it was formerly classified in the Liliaceae.

They grow from short rhizomes, and bear tufts of leaves, from which flowers emerge in racemes held on short stems above the leaves.

Cultivation and Uses

Some species such as *Ophiopogon japonicus* and *Ophiopogon planiscapus* are used as ground cover plants.

In Chinese medicine the tuber of *Ophiopogon japonicus*, known as mai men dong, is the cardinal herb for yin deficiency. According to the Chinese Herbal Medicine *Materia Medica*, the herb is sweet, slightly bitter and slightly cold, enters the heart, lung and stomach channels and nourishes the yin of the stomach, spleen, heart and lungs and clears heat and quiets irritability. It is used for hacking dry coughs, dry tongue and mouth and constipation. Liriope (genus) is used as a substitute.

2

Ornamental Crops under Protected Cultivation

Protected cultivation is future because Land and Water – Nature's gift to mankind is not unlimited and free forever. Everyday population is increasing and due to modernization the infrastructure work is increasing and day to day agricultural land is becoming less and less and at the same time the land value is increasing.

Indoor Ornamental Plants

Ficus Panda

Ficus is a genus of about 850 species of woody trees, shrubs, vines, epiphytes and hemiepiphytes in the family Moraceae. Collectively known as fig trees or figs, they are native throughout the tropics with a few species extending into the semi-warm temperate zone. The Common Fig (*F. carica*) is a temperate species native to southwest Asia and the Mediterranean region (from Afghanistan to Portugal), which has been widely cultivated from ancient times for its fruit, also referred to as figs. The fruit of most other species are also edible though they are usually of only local economic importance or eaten as bushfood. However, they are extremely important food resources for wildlife. Figs are also of considerable cultural importance throughout the tropics, both as objects of worship and for their many practical uses.

The Ficus Panda offered by us is a round, green, shiny plant usually found in gardens. Bowl shape of Ficus Panda provides it highly attractive look. Ficus Panda is a yellow coloured bushy plant offered by us at very reasonable rates.

Appearance:

- Bowl shaped
- Bushy
- Yellow colour

Ficus Tree Care: With more than 800 species available, ficus trees are one of the most popular decorative trees grown indoors. However, they are also rather sensitive plants, so get tips on how to care for them properly and help them thrive.

Caring for any houseplant starts long before the plant takes up residence in a kitchen, office or den. From properly choosing the best plant to understanding why a ficus tree loses leaves, these tips can help keep your plant healthy and vibrant.

Figure: *Ficus sycomorus*

Choosing a Ficus: Ficus trees are very temperamental and can become stressed from simple actions such as being moved from one room to another, so it is essential to choose a healthy specimen from the nursery before taking it home and immersing it in a completely new setting.

When choosing a ficus, look for the following characteristics.

- The tree should have a straight, sturdy trunk free from scars or warping.
- A ficus should also have rich green leaves with no signs of browning or yellowing, and no rips or tears.
- The soil should be moist, which indicates a well-watered plant.
- There should be no bugs in the leaves or soil.
- There should be no signs of recent leaf loss that would indicate plant stress.

Potting: Ficus trees have vigorous root growth and may easily become pot bound. Replanting the tree is easy, and should be done every one to two years depending on its growth. Choose a pot at least two inches wider in diametre than the previous pot to ensure plenty of space for the tree to grow. For many trees, this will need to be done soon after they have been purchased as they are moved to a new home.

Location:

- The tree should be placed in a location that receives some bright sunlight during the day, but it should not be in full sun all day long.
- The optimum temperature range is between 65 and 75 degrees Fahrenheit.
- Do not position the plant next to heaters, vents or windows that can create widely varied temperatures.

Watering and Humidity: Since ficus trees are native to more humid climates, they require careful watering and humidity control for the best growth.

- The soil should be kept moist but not wet.
- Place a saucer under the pot to provide extra drainage which will help prevent root rot.
- The tree should be watered after the top inch or two of soil has dried.
- Mist the tree gently once or twice daily, particularly while it is young or if it is adjusting to a new location. This can help it maintain its humidity.
- In extremely dry homes, adding a room humidifier can help adjust the humidity level to match the ficus tree's needs.

Leaf Loss: Leaf loss is typically the first sign of stress a ficus tree exhibits, but proper care can help minimize any problems. It is not uncommon for a tree to lose one-quarter of its leaves even after a simple environmental change such as switching rooms, but proper watering and temperature regulation can help mitigate additional stress. If the leaves turn yellow just prior to dropping off, it is a sign of over-watering, while dull leaves may indicate a lack of essential nutrients.

Fertilizing: Your ficus will need to be fed regularly to keep it healthy.

- Use a guaranteed analysis 10-10-10 slow release fertilizer.
- Apply it at half the strength of the average recommended dosage to reduce the chance of burning your plant.
- Apply it once a month during active growth, which is typically from spring through fall.

Seasonal Care: Ficus trees are seasonal plants and have different growing cycles for different times of the year.

- During spring and summer, the trees grow quicker and require more water, fertilizer and sunlight to stay healthy.
- When nighttime outdoor temperatures remain above 50 degrees Fahrenheit, these trees may be placed outdoors for the summer in a bright, sunny area where they can take advantage of natural humidity and rainwater.
- In late autumn and winter, as the days grow colder, the growth slows down. Less water and fertilizer are required during cooler seasons.

Additional Ficus Care Advice: For the healthiest, most beautiful ficus trees, remember these additional tips:

- Prune and shape the trees in early spring just as the growing season begins.
- Ficus trees can be groomed into bonsai if desired, using both buried and exposed root designs.
- Wipe dust off the tree's leaves regularly to keep the pores clear so they can absorb humidity more easily.
- Do not use wax-based cleaners on the leaves because it clogs the pores.

Common Fig

The common fig (*Ficus carica*) is a species of flowering plant in the genus *Ficus*, from the family Moraceae, known as the common fig (or just the fig), *anjeer* (Hindi), and *dumur* (Bengali). It is the source of the fruit also called the fig, and as such is an important crop in those areas where it is grown commercially. Native to the Middle East and western Asia, it has been sought out and cultivated by man since ancient times, and is now widely grown throughout the temperate world, both for its fruit and as an ornamental plant.

Description: It is a gynodioecious (functionally dioecious), deciduous tree or large shrub, growing to a height of 7–10 metres (23–33 ft), with smooth white bark. Its fragrant leaves are 12–25 centimetres (4.7–9.8 in) long and 10–18 centimetres (3.9–7.1 in) across, and deeply lobed with three or five lobes. The complex inflorescence consists of a hollow fleshy structure called the syconium, which is lined with numerous unisexual flowers. The flower itself is not visible outwardly, as it blooms inside the infructescence. Although commonly referred to as a fruit, the fig is actually the infructescence or scion of the tree, known as a false fruit or multiple fruit, in which the flowers and seeds are borne. It is a hollow-ended stem containing many flowers. The small orifice (ostiole) visible on the middle of the fruit is a narrow passage, which allows the specialized fig wasp *Blastophaga psenes* to enter the fruit and pollinate the flower, whereafter the fruit grows seeds.

The edible fruit consists of the mature syconium containing numerous one-seeded fruits (druplets). The fruit is 3–5 centimetres (1.2–2.0 in) long, with a green skin, sometimes ripening towards purple or brown. *Ficus carica* has milky sap (laticifer). The sap of the fig's green parts is an irritant to human skin.

Figure: *Variegated fig*

Habitat: The common fig tree has been cultivated since ancient times and grows wild in dry and sunny areas, with deep and fresh soil;

also in rocky areas, from sea level to 1,700 metres. It prefers light and medium soils, requires well-drained soil, and can grow in nutritionally poor soil. Like all fig trees, *Ficus carica* requires wasp pollination of a particular species of wasp (*Blastophaga psenes*) to produce seeds. The plant can tolerate seasonal drought, and the Middle Eastern and Mediterranean climate is especially suitable for the plant. Situated in a favourable habitat, old specimens when mature can reach a considerable size and form a large dense shade tree. Its aggressive root system precludes its use in many urban areas of cities, but in nature helps the plant to take root in the most inhospitable areas. Common fig tree is mostly a phreatophyte that lives in areas with standing or running water, grows well in the valleys of the rivers and ravines saving no water, having strong need of water that is extracted from the ground.

The deep-rooted plant searches groundwater, in aquifers, ravines, or cracks in the rocks. The fig tree, with the water, cools the environment in hot places, creating a fresh and pleasant habitat for many animals that take shelter in its shade in the times of intense heat.

Ecology: *Ficus carica* is dispersed by birds and mammals that scatter their seeds in droppings. Fig fruit is an important food source for much of the fauna in some areas, and the tree owes its expansion to those that feed on its fruit. The common fig tree also sprouts from the root and stolon issues.

The infructescence is pollinated by a symbiosis with a kind of fig wasp (*Blastophaga psenes*). The fertilized female wasp enters the fig through the scion, which is a tiny hole in the crown (the ostiole). She crawls on the inflorescence inside the fig and pollinates some of the female flowers. She lays her eggs inside some of the flowers and dies. After weeks of development in their galls, the male wasps emerge before females through holes they produce by chewing the galls. The male wasps then fertilize the females by depositing semen in the hole in the gall. The males later return to the females and enlarge the holes to enable the females to emerge. Then some males enlarge holes in the scion, which enables females to disperse after collecting pollen from the developed male flowers. Females have a short time (<48 hours) to find another fig tree with receptive scions to spread the pollen, assist the tree in reproduction, and lay their own eggs to start a new cycle.

History: The edible fig is one of the first plants that was cultivated by humans. Nine subfossil figs of a parthenocarpic (and therefore sterile) type dating to about 9400–9200 BC were found in the early Neolithic village Gilgal I (in the Jordan Valley, 13 km north of Jericho).

The find predates the domestication of wheat, barley, and legumes, and may thus be the first known instance of agriculture. It is proposed that this sterile but gustatively desirable type was planted and cultivated intentionally, one thousand years before the next crops were domesticated (wheat and rye).

Figs were also a common food source for the Romans. Cato the Elder, in his *De Agri Cultura*, lists several strains of figs grown at the time he wrote his handbook: the Mariscan, African, Herculanean, Saguntine, and the black Tellanian (*De agri cultura*, ch. 8). The fruits were used, among other things, to fatten geese for the production of a precursor of foie gras.

It was cultivated from Afghanistan to Portugal, also grown in Pithoragarh in the Kumaon hills of India. From the 15th century onwards, it was grown in areas including Northern Europe and the New World. In the 16th century, Cardinal Reginald Pole introduced fig trees to Lambeth Palace in London.

Cultivation: The common fig is grown for its edible fruit throughout the temperate world. It is also grown as an ornamental tree, and the cultivar 'Brown Turkey' has gained the Royal Horticultural Society's Award of Garden Merit. Figs can be found in continental climates with hot summers as far north as Hungary and Moravia, and can be harvested up to four times per year. Thousands of cultivars, most named, have been developed as human migration brought the fig to many places outside its natural range.

Two crops of figs are potentially produced each year. The first or breba crop develops in the spring on last year's shoot growth. In contrast, the main fig crop develops on the current year's shoot growth and ripens in the late summer or fall. The main crop is generally superior in both quantity and quality to the breba crop. However, some cultivars produce good breba crops (e.g., 'Black Mission', 'Croisic', and 'Ventura').

There are basically three types of edible figs:

- Persistent (or common) figs have all female flowers that do not need pollination for fruiting; the fruit can develop through parthenocarpic means. This is a popular horticulture fig for home gardeners. Dottato (Kadota), Black Mission, Brown Turkey, Brunswick, and Celeste are some representative cultivars.
- Caducous (or Smyrna) figs require cross pollination by the fig wasp with pollen from caprifigs for the fruit to mature. If not pollinated the immature fruits drop. Some cultivars are Smyrne (Lob Incir in Turkey) - (Calimyrna in the Great Central Valley USA), Marabout, Inchàrio, and Zidi.

- Intermediate (or San Pedro) figs set an unpollinated breba crop, but need pollination for the later main crop. Examples are Lampeira, King, and San Pedro.

The fig likes dry sunny sites, the soil dry or drained. Excessive growth has to be limited to promote the fruiting. It thrives in both sandy and rocky soil. As the sun is really important it is better to avoid shades. Some varieties are more adapted to harsh and wet climates.

Propagation: Figs plants are easy to propagate through several methods. Propagation using seeds is not the preferred method since vegetative methods exist that are quicker and more reliable, that is, they do not yield the inedible caprifigs. However, those desiring to can plant seeds of dried figs with moist sphagnum moss or other media in a zip lock bag and expect germination in a few weeks to several months. The tiny plants can be transplanted out little by little once the leaves open, and despite the tiny initial size can grow to 1 foot (30 cm) or more within one year from planting seeds.

Main vegetative propagation, or spring propagation: before the tree starts growth, cut 15–25 cm (6–10 inch) shoots that have healthy buds at their ends, and set into a moist mix of soil and peat-moss located in shade in first time, buried 3/4 of their length. Larger diametre stems are better – intermediate cuttings on branches can be done too (up to diam. 3/4") – but in this case the upper side must be cut inclined, thus marking the upper part, to avoid planting upside-down. Grow one year in a nursery, in a pot or in-ground spaced one foot apart, till winter. Before the plant starts growth, plant it in the desired final location.

For propagation in the mid-summer months, air layer new growth in August (mid-summer) or insert hardened off 15–25 cm (6-10 inches) shoots into moist perlite or a sandy soil mix, keeping the cuttings shaded until new growth begins; then gradually move them into full sun. For spring propagation, before the tree starts growth, cut 15–25 cm (6-10 inches) shoots that have healthy buds at their ends, and set into a moist perlite and/or sandy soil mix located in the shade. Once the cuttings start to produce leaves, bury them up to the bottom leaf to give the plant a good start in the desired location.

An alternative propagation method is bending over a taller branch, scratching the bark to reveal the green inner bark, then pinning the scratched area tightly to the ground. Within a few weeks, roots will develop and the branch can be clipped from the mother plant and transplanted where desired.

Nutrition: Figs are among the richest plant sources of calcium and fibre. According to USDA data for the Mission variety, dried figs are richest in fibre, copper, manganese, magnesium, potassium, calcium, and vitamin K, relative to human needs. They have smaller amounts of many other nutrients. Figs have a laxative effect and contain many antioxidants. They are a good source of flavonoids and polyphenols including gallic acid, chlorogenic acid, syringic acid, (+)-catechin, (–)-epicatechin and rutin. In one study, a 40-gram portion of dried figs (two medium size figs) produced a significant increase in plasma antioxidant capacity.

Figure: *"Schiocca": calabrian dried figs*

Cultural Aspects: In the Book of Genesis in the Bible, Adam and Eve clad themselves with fig leaves (Genesis 3:7) after eating the "forbidden fruit" from the Tree of Knowledge of Good and Evil. Likewise, fig leaves, or depictions of fig leaves, have long been used to cover the genitals of nude figures in painting and sculpture. Art collectors and exhibitors often added these depictions long after the original work was completed. The use of the fig leaf as a protector of modesty or shield of some kind has entered the language.

The Book of Deuteronomy specifies the fig as one of the Seven Species (Deuteronomy 8:7-8), describing the fertility of the land of Canaan. This is a set of seven plants indigenous to the Middle East that together can provide food all year round. The list is organized by date of harvest, with the fig being fourth due to its main crop ripening during summer.

Also in the Bible (Matthew 21:18–22 and Mark 11:12–14, 19–21) is a story of Jesus finding a fig tree when he was hungry; the tree only had leaves on it, but no fruit. Jesus, then, curses the fig tree, which withers.

The biblical quote "each man under his own vine and fig tree" (1 Kings 4:25) has been used to denote peace and prosperity. It was commonly quoted to refer to the life that would be led by settlers in the American West, and was used by Theodor Herzl in his depiction of the future Jewish Homeland: "We are a commonwealth. In form it is new, but in purpose very ancient. Our aim is mentioned in the First Book of Kings: 'Judah and Israel shall dwell securely, each man under his own vine and fig tree, from Dan to Beersheba". US President George Washington, writing in 1790 to the Touro Synagogue of Newport, Rhode Island, extended the metaphor to denote the equality of all Americans regardless of faith.

Buddha achieved enlightenment under the bodhi tree, a large and old sacred fig tree (Ficus religiosa, or Pipal).

Sura 95 of the Qur'an is named *al-Tîn* (Arabic for "The Fig"), as it opens with the oath "By the fig and the olive." The fruit is also mentioned elsewhere in the Qur'an. Within the Hadith, Sahih al-Bukhari records Prophet Muhammad stating: "If I had to mention a fruit that descended from paradise, I would say this is it because the paradisiacal fruits do not have pits...eat from these fruits for they prevent hemorrhoids, prevent piles and help gout."

In Greek mythology, the god Apollo sends a crow to collect water from a stream for him. The crow sees a fig tree and waits for the figs to ripen, tempted by the fruit. He knows that he is late and that his tardiness will be punished, so he gets a snake from the stream and collects the water. He presents Apollo with the water and uses the snake as an excuse. Apollo sees through the crow's lie and throws the crow, goblet, and snake into the sky where they form the constellations Hydra, Crater, and Corvus.

In Aristophanes' *Lysistrata* one of the women boasts about the "curriculum" of initiation rites she went through to become an adult woman (*Lys.* 641–7). As her final accomplishment before marriage, when she was already a fair girl, she bore the basket as a kanephoros, wearing a necklace of dried figs.

Aglonima Hybrid Spp

The aglaonema is a highly decorative plant with several interesting varieties. There seems to be a little controversy in the books over whether this plant is easy or difficult to grow. The simple rule to follow is this: the lighter the variegation, the more light it needs. Otherwise, these are slow growing, dependably attractive and make excellent foliage plants. They have large, narrow oval leaves on short stems.

Growing Conditions:

Light: The darker green varieties can grow in near shade, while the variegated varieties require brighter light. Do not expose to direct sun.

Water: Water thoroughly in the summer, and mist often to raise humidity. During the winter, reduce watering but do not let the plant dry completely. Seek to raise humidity in general.

Temperature: They do not like cold drafts or temperatures below 65°F. The warmer, the better.

Soil: A well-drained potting soil is perfect.

Fertilizer: Slow-release pellets or liquid fertilizer during the growing season.

Propagation

Aglaonema are not typically propagated by home growers, but they can be divided during repotting. Small shoots can be potted as individual plants.

Repotting: They are slow growing and will only need repotting every other year. Aglaonema are generally low-growing plants that will only very gradually reveal their trunks.

Varieties: Aglaonema have been hybridized to produce interesting variegated leaves. The A. commutatum in widely available, in both the Silver Spear (variegated) form and a green form. A pure green A. modestum is seen more rarely. Beware the fruit of the A. crispum, which are toxic. Other variegated forms include A. pictum, A. silver queen (almost totally silver), and A. pseudobracteatum.

Grower's Tips: Because of their high humidity requirements, some growers consider aglaonema to be greenhouse plants. It's true they will do best in the warm, humid and bright environment of a greenhouse, but they can successfully be grown indoors by coming as close as possible to these conditions. The number one rule to remember with aglaonema is this: keep them warm and moist. If you do this, you'll be rewarded with a long-lasting, stable houseplant that will not soon outgrow its pot.

Dracaena Braunii

Dracaena braunii is a species of the genus *Dracaena*. It is also known as *Dracaena sanderiana*, after the German–English gardener Henry Frederick Conrad Sander (1847–1920). Common names include *Sander's dracaena, ribbon dracaena, lucky bamboo, curly bamboo, Chinese water bamboo, friendship bamboo, Goddess of Mercy plant, Belgian evergreen*, and *ribbon plant*.

D. braunii is one of a group of small, shrub by species with slender stems and flexible strap-shaped leaves that grow as understorey plants in rainforests. It is native to Cameroon in tropical west Africa. It is an upright shrub growing to 1.5 metres (5 ft) tall, with leaves 15–25 cm (6–10 in) long and 1.5–4 cm (1–2 in) broad at the base.

Cultivation and Uses: *Dracaena braunii* and related species are popular houseplants, with numerous cultivars sold. It can survive in many indoor conditions, but indirect lighting is best as direct sunlight can cause the leaves to turn yellow and burn.

Although it grows better in soil, it is often sold with the roots in water. The water should be completely changed every two weeks. The water should be bottled water, soft tap water with very little fluoride, or even water from a filtered, established aquarium. It does best in bright, indirect lighting and temperatures from 15 to 25 °C (59 to 77 °F).

Yellow or brown leaf edges may be caused by too much direct light, crowded roots, or fluoridated or chlorinated water, the latter of which can no longer be prevented by leaving tap water exposed to the air as chloramine has largely replaced the more volatile chlorine in water treatment. Salty or softened water can also cause this.

Twisted shapes can be produced by rotating the plant with respect to gravity and directed light sources. This is difficult to achieve for most home users, but not impossible with a lot of spare time and a lot of patience.

Often in large chain pet shops it will be sold as an aquatic plant. While it will live for months like this, it will eventually rot unless the sprouts are allowed to grow above the surface.

Other Information:

- *Dracaena braunii* is toxic to pets.
- *Dracaena braunii* can flower in autumn, winter, and early spring.

Lucky Bamboo

It is marketed in the developed world as a Chinese decorative plant "lucky bamboo" (although unrelated to bamboo and not native to Asia), propagated from short cuttings, usually in water. *Dracaena braunii* has long been associated with the Eastern practice of feng shui. Lucky bamboo is believed to be an ideal example of the thriving wood and water element, with the addition of a red ribbon sometimes tied around the stalks which is believed to "fire" the positive flow of energy or chi in the room. The number of stalks also has meaning: three for happiness; five for wealth; six for health. (Four stalks are avoided since the word 'four' in Chinese sounds too similar to the word for 'death'.)

Keeping *lucky bamboo* inside house and business places is believed to bring happiness and prosperity and this belief has promoted a huge sales of *lucky bamboo* shoots grown in decorative pots. In India, the plants are usually imported from China and Taiwan and resold in attractive pots and the plant has become most popular indoor plant in certain parts of India.

Cordyline Terminalis

Cordyline fruticosa is an evergreen flowering plant in the Asparagus family, Asparagaceae, known by a wide variety of common names including Cabbage Palm, Good Luck Plant, Palm Lily, Ti Plant.

Formerly treated in the families Agavaceae and Laxmanniaceae (now both subfamilies of the Asparagaceae in the APG III system), it is a woody plant growing up to 4 m (13 ft) tall, with leaves 30–60 cm (12–24 in) (rarely 75 cm or 30 in) long and 5–10-centimetre (2.0–3.9 in) wide at the top of a woody stem. It produces 40–60-centimetre (16–24 in) long panicles of small scented yellowish to red flowers that mature into red berries.

It is native to tropical southeastern Asia, Papua New Guinea, Melanesia, northeastern Australia, the Indian Ocean, and parts of Polynesia. It is not native to either Hawaii or New Zealand but was introduced to both by Polynesian settlers.

Cultivation and Uses: The species was spread from its native range throughout Polynesia as a cultivated plant. Its starchy rhizomes, which are very sweet when the plant is mature, were eaten as food or as medicine, and its leaves were used to thatch the roofs of houses, and to wrap and store food. The plant or its roots are referred to in most Polynesian languages as *tî*. Mâori ranked the sweetness of the plant above the other *Cordyline* species native to New Zealand.

Leaves were also used to make items of clothing including skirts worn in dance performances. The Hawaiian hula skirt is a dense skirt with an opaque layer of at least 50 green leaves and the bottom (top of the leaves) shaved flat. The Tongan dance dress, the *sisi*, is an apron of about 20 leaves, worn over a tupenu, and decorated with some yellow or red leaves.

In Vanuatu, *Cordyline* leaves, known locally by the Bislama name *nanggaria*, are worn tucked into a belt in traditional dances, with different varieties having particular symbolic meanings. Cordylines are often planted outside nakamals.

In ancient Hawai the plant was thought to have great spiritual power; only kahuna (high priests) and ali'i (chiefs) were able to wear leaves around their necks during certain ritual activities. *Tî* leaves were also used to make lei, and to outline borders between properties it was also planted at the corners of the home to keep ghosts from entering the home or property (for which its alternative name: *terminalis)*. To this day some Hawaiians plant *tî* near their houses to bring good luck. The leaves are also used for lava sledding. A number of leaves are lashed together and people ride down hills on them.

The roots of the *tî* plant were used as a glossy covering on surfboards in Hawaii in the early 1900s.

Ti is a popular ornamental plant, with numerous cultivars available, many of them selected for green or reddish or purple foliage.

In Hawaii, *tî* rhizomes are fermented and distilled to make okolehao, a liquor.

Drecena

Dracaena is a genus of about 40 species of trees and succulent shrubs. In the APG III classification system, it is placed in the family Asparagaceae, subfamily Nolinoideae (formerly the family Ruscaceae). It has also formerly been separated (sometimes with *Cordyline*) into the family Dracaenaceae or placed in the Agavaceae (now Agavoideae). The majority of the species are native to Africa, with a few in southern Asia and one in tropical Central America. The segregate genus *Pleomele* is now generally included in *Dracaena*. The genus *Sanseviera* is closely related, and has recently been synonymized under *Dracaena* in the Kubitzki system.

Species of *Dracaena* have a secondary thickening meristem in their trunk, which is quite different from the thickening meristem found in dicotyledonous plants and is termed dracaenoid thickening by some authors. This characteristic is shared with members of the Agavoideae and Xanthorrhoeoideae among other members of the Asparagales.

D. americana, *D. arborea*, *D. cinnabari*, *D. draco*, *D. ombet*, and *D. tamaranae* are commonly known as dragon trees and grow in arid semi-desert areas. They are tree-sized with stout trunks and stiff, broad-based leaves. The remaining species are known collectively as shrubby dracaenas. They are smaller and shrub-like, with slender stems and flexible strap-shaped leaves, and grow as understorey plants in rainforests.

Uses: *Ornamental:* Some shrubby species, such as *D. deremensis*, *D. fragrans*, *D. godseffiana*, *D. marginata*, and *D. braunii*, are popular as houseplants. Many of these are toxic to pets, though not humans, according to the ASPCA among others. Rooted stem cuttings of *D. braunii* are widely marketed in the U.S.A. and the UK as "lucky bamboo", although only superficially resembling true bamboos.

Other Uses: A bright red resin, dragon's blood, is produced from *D. draco* and, in ancient times, from *D. cinnabari*. Modern dragon's blood is however more likely to be from the unrelated *Daemonorops* rattan palms.

Greenhouse

A greenhouse (also called a glasshouse) is a building or complex in which plants are grown. These structures range in size from small sheds to industrial-sized buildings. A miniature greenhouse is known as a cold frame. Commercial glass greenhouses are often high tech production facilities for vegetables or flowers. The glass greenhouses are filled with equipment like screening installations, heating, cooling, lighting and also may be automatically controlled by a computer to maximize potential growth.

A greenhouse is a structural building with different types of covering materials, such as a glass or plastic roof and frequently glass or plastic walls; it heats up because incoming visible sunshine is absorbed inside the structure. Air warmed by the heat from warmed interior surfaces is retained in the building by the roof and wall; the air that is warmed near the ground is prevented from rising indefinitely and flowing away. This is not the same mechanism as the "greenhouse effect".

Types

Greenhouses can be divided into glass greenhouses and plastic greenhouses. Plastics mostly used are polyethylene film and multiwall sheets of polycarbonate material, or PMMA acrylic glass. Commercial glass greenhouses are often high-tech production facilities for vegetables or flowers. The glass greenhouses are filled with equipment such as screening installations, heating, cooling and lighting, and may be automatically controlled by a computer.

Uses

Greenhouses allow for greater control over the growing environment of plants. Depending upon the technical specification of a greenhouse, key factors which may be controlled include temperature, levels of light and shade, irrigation, fertilizer application, and atmospheric humidity. Greenhouses may be used to overcome shortcomings in the growing qualities of a piece of land, such as a short growing season or poor light levels, and they can thereby improve food production in marginal environments.

As they may enable certain crops to be grown throughout the year, greenhouses are increasingly important in the food supply of high-latitude countries. One of the largest complexes in the world is in Almería, Andalucía, Spain, where greenhouses cover almost 50,000 acres (200 km^2)

Greenhouses are often used for growing flowers, vegetables, fruits, and transplants. Special greenhouse varieties of certain crops, such as tomatoes, are generally used for commercial production. Many vegetables and flowers can be grown in greenhouses in late winter and early spring, and then transplanted outside as the weather warms. Bumblebees are the pollinators of choice for most pollination, although other types of bees have been used, as well as artificial pollination. Hydroponics can be used to make the most use of the interior space.

The relatively closed environment of a greenhouse has its own unique management requirements, compared with outdoor production. Pests and diseases, and extremes of heat and humidity, have to be controlled, and irrigation is necessary to provide water. Most greenhouses use sprinklers or drip lines. Significant inputs of heat and light may be required, particularly with winter production of warm-weather vegetables.

Greenhouses also have applications outside of the agriculture industry. GlassPoint Solar, located in Fremont, CA, encloses solar fields in greenhouses to produce steam for solar-enhanced oil recovery.

History

The idea of growing plants in environmentally controlled areas has existed since Roman times. The Roman emperor Tiberius ate a cucumber-like vegetable daily. The Roman gardeners used artificial methods (similar to the greenhouse system) of growing to have it available for his table every day of the year. Cucumbers were planted in wheeled carts which were put in the sun daily, then taken inside to keep them warm at night. The cucumbers were stored under frames or in cucumber houses glazed with either oiled cloth known as *specularia* or with sheets of selenite (a.k.a. *lapis specularis*), according to the description by Pliny the Elder.

In the 13th century, greenhouses were built in Italy to house the exotic plants that explorers brought back from the tropics. They were originally called *giardini botanici* (botanical gardens).

'Active' greenhouses, in which it is possible for the temperature to be increased or decreased manually, appeared much later. Sanga yorok written in the year 1450 AD in Korea, contained descriptions of a greenhouse, which was designed to regulate the temperature and humidity requirements of plants and crops. One of the earliest records of the Annals of the Joseon Dynasty in 1438 confirms growing mandarin trees in a Korean traditional greenhouse during the winter and installing a heating system of Ondol.

The concept of greenhouses also appeared in Netherlands and then England in the 17th century, along with the plants. Some of these early attempts required enormous amounts of work to close up at night or to winterize. There were serious problems with providing adequate and balanced heat in these early greenhouses. Today, the Netherlands has many of the largest greenhouses in the world, some of them so vast that they are able to produce millions of vegetables every year.

The French botanist Charles Lucien Bonaparte is often credited with building the first practical modern greenhouse in Leiden, Holland during the 1800s to grow medicinal tropical plants. Originally only on the estates of the rich, the growth of the science of botany caused greenhouses to spread to the universities. The French called their first greenhouses *orangeries*, since they were used to protect orange trees from freezing. As pineapples became popular, *pineries*, or pineapple pits, were built.

Experimentation with the design of greenhouses continued during the 17th century in Europe, as technology produced better glass and construction techniques improved. The greenhouse at the Palace of Versailles was an example of their size and elaborateness; it was more than 500 feet (150 m) long, 42 feet (13 m) wide, and 45 feet (14 m) high.

The golden era of the greenhouse was in England during the Victorian era, where the largest glasshouses yet conceived were constructed, as the wealthy upper class and aspiring botanists competed to build the most elaborate buildings. A good example of this trend is the pioneering Kew Gardens. Joseph Paxton, who had experimented with glass and iron in the creation of large greenhouses as the head gardener at Chatsworth, in Derbyshire, working for the Duke of Devonshire, designed and built The Crystal Palace in London, (although the latter was constructed for both horticultural and non-horticultural exhibition).

Other large greenhouses built in the 19th century, included the New York Crystal Palace, Munich's Glaspalast and the Royal Greenhouses of Laeken (1874–1895) for King Leopold II of Belgium.

In Japan, the first greenhouse was built in 1880 by Samuel Cocking, a British merchant who exported herbs.

In the 20th century, the geodesic dome was added to the many types of greenhouses. Notable examples are the Eden Project, in Cornwall, The Rodale Institute in Pennsylvania, the Climatron at the Missouri Botanical Garden in St. Louis, Missouri, and Toyota Motor Manufacturing Kentucky.

Greenhouse structures adapted in the 1960s when wider sheets of polyethylene film became widely available. Hoop houses were made by several companies and were also frequently made by the growers themselves. Constructed of aluminum extrusions, special galvanized steel tubing, or even just lengths of steel or PVC water pipe, construction costs were greatly reduced. This resulted in many more greenhouses being constructed on smaller farms and garden centres. Polyethylene film durability increased greatly when more effective UV-inhibitors were developed and added in the 1970s; these extended the usable life of the film from one or two years up to 3 and eventually 4 or more years.

Gutter-connected greenhouses became more prevalent in the 1980s and 1990s. These greenhouses have two or more bays connected by a common wall, or row of support posts. Heating inputs were reduced as the ratio of floor area to roof area was increased substantially. Gutter-connected greenhouses are now commonly used both in production and in situations where plants are grown and sold to the public as well. Gutter-connected greenhouses are commonly covered with structured polycarbonate materials, or a double layer of polyethylene film with air blown between to provide increased heating efficiencies.

Greenhouse Ventilation

Ventilation is one of the most important components in a successful greenhouse. If there is no proper ventilation, greenhouses and their growing plants can become prone to problems. The main purposes of ventilation are to regulate the temperature and humidity to the optimal level, and to ensure movement of air and thus prevent build-up of plant pathogens (such as *Botrytis cinerea*) that prefer still air conditions. Ventilation also ensures a supply of fresh air for photosynthesis and plant respiration, and may enable important pollinators to access the greenhouse crop.

Ventilation can be achieved via use of vents - often controlled automatically via a computer - and recirculation fans.

Greenhouse Heating

Heating or electricity is one of the most considerable costs in the operation of greenhouses across the globe, especially in colder climates. The main problem with heating a greenhouse as opposed to a building that has solid opaque walls is the amount of heat lost through the greenhouse covering. Since the coverings need to allow light to filter into the structure, they conversely cannot insulate very well. With traditional plastic greenhouse coverings having an R-value of around 2, a great amount of money is therefore spent to continually replace the heat lost. Most greenhouses, when supplemental heat is needed use natural gas or electric furnaces.

Passive heating methods exist which seek heat using low energy input. Solar energy can be captured from periods of relative abundance (day time/summer), and released to boost the temperature during cooler periods (night time/winter). Waste heat from livestock can also be used to heat greenhouses; e.g. placing a chicken coop inside a greenhouse recovers the heat generated by the chickens, which would otherwise be wasted.

Electronic controllers are often used to monitor the temperature and adjusts the furnace operation to the conditions. This can be as simple as a basic thermostat, but can be more complicated in larger greenhouse operations.

Greenhouse Carbon Dioxide Enrichment

The possibility of using carbon dioxide enrichment in greenhouse cultivation to enhance plant growth has been known for nearly 100 years. After the development of equipment for the controlled serial enrichment of carbon dioxide, the technique was established on a broad scale in the Netherlands. Secondary metabolites, e. g. cardiac glycosides in Digitalis lanata, are produced in higher amounts by greenhouse cultivation at enhanced temperature and at enhanced carbon dioxide concentration. Commercial greenhouses are now frequently located near appropriate industrial facilities for mutual benefit. For example, Cornerways Nursery in the UK is strategically placed near a major sugar refinery, consuming both waste heat and CO_2 from the refinery which would otherwise be vented to atmosphere. The refinery reduces its carbon emissions, whilst the nursery enjoys boosted tomato yields and does not need to provide its own greenhouse heating.

Bioshelter

A bioshelter is a solar greenhouse managed as an indoor ecosystem. The word bioshelter was coined by the New Alchemy Institute and solar designers Sean Wellesley-Miller and Day Chahroudi. The term was created to distinguish their work in greenhouse design and management from twentieth century petro-chemical fuelled monoculture greenhouses.

New Alchemy's pioneering work in ecological design is documented in their published Journals and Reports. In 1976 the Alchemists built the Cape Cod Ark bioshelter and her sister The Prince Edward Island Ark. For the next 15 years the New Alchemy Institute studied and reported on the use of these prototype food producing ecosystems.

A bioshelter (life-shelter) involves two fields of knowledge and design. The first is architecture designed to nurture an ecosystem within. A bioshelter structure uses glazing to contain and protect the living biology inside, control air exchange and absorb energy. The building exchanges nutrients, gases and energy with the surrounding environment, produces crops, and recycles waste organic material into the soil. Solar energy is stored as heat energy in thermal mass such as water, stone, masonry, soil and plant biomass.

The second is the biology inside the bioshelter. Earle Barnhart of the New Alchemy Institute has compared a bioshelter to a contained ecosystem. Solar heat is absorbed and stored in thermal mass to moderate air temperatures and provide heat for later use. Water moves from rainfall to fishponds to soil to plants and finally to water vapor. Year-round habitat is provided for beneficial insects . Ecological relationships between pests and their predators reduce the number of pests. Gases are exchanged among the animals, insects, micro-organisms, soil and plants. Nutrient cycles are developed between fish, plant & soil. Within the bioshelter are a variety of microclimates. The south areas receive the most direct sunlight. The east and west areas can be shaded for a portion of the day. Higher levels in a growing space will be warmer. A well-designed bioshelter, managed by human intelligence, can shelter a community of people, food crops, edible fish, and a diverse ecosystem of plants, animals and soil life.

Controlled Ecological Life Support System

Controlled (or Closed) Ecological Life Support Systems (acronym CELSS) are a type of scientific endeavour to create a self-supporting life support system for space stations and colonies typically through controlled closed ecological systems, such as the BioHome, BIOS-3, Biosphere 2, and Mars Desert Research Station.

CELSS was first pioneered by the Soviet Union during the famed "Space Race" in the 1950s–60s. Originated by Konstantin Tsiolkovsky and furthered by V.I. Vernadsky, the first forays into this science were the use of closed, unmanned ecosystems, expanding into the research facility known as the BIOS-3.

Then in 1965, manned experiments began in the BIOS-3.

Rationale

Space travel, thus far, has been limited to our own Earth–Moon system. Also, everything that astronauts would need in the way of life support (air, water, and food) has been brought with them. This may be economical for short missions of spacecraft, but it is not the most elegant solution when dealing with the life support systems of a long-term craft (such as a generation ship) or a settlement.

The aim of CELSS is to create a regenerative environment that can support and maintain human life via agricultural means.

Components of CELSS

Air Revitalization: In non-CELSS environments, air replenishment and CO_2 processing typically consists of stored air tanks and CO_2 scrubbers. The drawback to this method lies in the fact that upon depletion the tanks would have to be refilled; the scrubbers would also require replacement after they become ineffective.

There is also the issue of processing toxic fumes, which come from the synthetic materials used in the construction of habitats. Therefore the issue of how air quality is maintained requires attention; in experiments, it was found that the plants also removed volatile organic compounds offgassed by synthetic materials used thus far to build and maintain all man-made habitats.

In CELSS, air is initially supplied by external supply, but is maintained by the use of foliage plants, which create oxygen in photosynthesis (aided by the waste-byproduct of human respiration, CO_2). Eventually, the main goal of a CELSS environment is to have foliage plants take over the complete and total production of oxygen needs; this would make the system a *closed*, instead of *controlled*, system.

Food/consumables Production

As with all present forays into space, crews have had to store all consumables they require prior to launch. Typically, hard-food consumables were freeze dried so that the craft's weight could be reduced. Of course, in a self-sustaining ecosystem, a place for crops to grow would be set aside, allowing foods to be grown and cultivated. The larger the group of people, the more crops would have to be grown.

As for water, experiments have shown that it would be derived from condensate in the air (a byproduct of air conditioning and vapors), as well as excess moisture from plants. It would then have to be filtered by some means, either by nature or by machine.

Waste-water Treatment

Early space-flight had travellers either ejecting their wastes into space or storing it for a return trip.

CELSS studied means of breaking down human wastes and, if possible, integrating the processed products back into the ecology. For instance, urine was processed into water, which was safe for use in toilets and watering plants.

Wastewater treatment makes use of plants, particularly aquatic, to process the wastewater. It has been shown that the more waste is treated by the aquatic plants (or, more specifically, their root systems), the larger the aquatic plants grow.

In tests, such as those done in the BioHome, the plants also made viable compost as a growth medium for crops.

Closed versus Controlled

Closed systems are totally self-reliant, recycling everything indefinitely with no external interaction. The life of such a system is limited, as the entropy of a closed system can only increase with time. But if the otherwise closed system is allowed to accept high-temperature radiant energy from an external source (e.g., sunlight) and to reject low-temperature waste heat to deep space, it can continue indefinitely. An example of such a system is the Earth itself.

Controlled systems, by contrast, depend on certain external interactions such as periodic maintenance.

Vertical Farming

Vertical farming is cultivating plant or animal life within a skyscraper greenhouse or on vertically inclined surfaces. The modern idea of vertical farming uses techniques similar to glass houses, where natural sunlight can be augmented with artificial lighting.

Types

"Vertical farming" was coined by Gilbert Ellis Bailey in 1915 in his book *Vertical Farming*. This was not the current meaning—he wrote about farming underground with the use of explosives. Modern usage refers to skyscrapers using some degree of natural light.

Mixed-use Skyscrapers

Mixed-use skyscrapers were proposed and built by architect Ken Yeang. Yeang proposes that instead of hermetically sealed mass-produced agriculture that plant life should be cultivated within open air, mixed-use skyscrapers for climate control and consumption (i.e. a personal or communal planting space as per the needs of the individual). This version of vertical farming is based upon personal or community use rather than the wholesale production and distribution plant and animal life that aspires to feed an entire city. It thus requires less of an initial investment than Despommier's "The Vertical Farm". However, neither Despommier nor Yeang are the conceptual "originators", nor is Yeang the inventor of vertical farming in skyscrapers.

Despommier's Skyscrapers

Ecologist Dickson Despommier argues that vertical farming is legitimate for environmental reasons. He claims that the cultivation of plant and animal life within skyscrapers will produce less embedded energy and toxicity than plant and animal life produced on natural landscapes.

He moreover claims that natural landscapes are too toxic for natural, agricultural production, despite the ecological and environmental costs of extracting materials to build skyscrapers for the simple purpose of agricultural production.

Vertical farming according to Despommier thus discounts the value of natural landscape in exchange for the idea of "skyscraper as spaceship". Plant and animal life are mass-produced within hermetically sealed, artificial environments that have little to do with the outside world. In this sense, they could be built anywhere regardless of the context. This is unlikely to be advantageous with regards to energy consumption as the internal environment must be maintained to sustain life within the skyscraper. However, this is not necessarily the case, as one of the most important features of a vertical farm is that it would contain some form of renewable energy technology, be it solar panels, wind turbines, or a water capture system, and could contain all three. The vertical farm is designed to be sustainable, and to enable nearby inhabitants to work at the farm.

Despommier's concept of "The Vertical Farm" emerged in 1999 at Columbia University. It promotes the mass cultivation of plant and animal life for commercial purposes in skyscrapers.

History

A commercial high-rise farm such as 'The Vertical Farm' has never been built, yet extensive photographic documentation and several historical books on the subject suggest that research on the subject was not diligently pursued. New sources indicate that a tower hydroponicum existed in Armenia prior to 1951.

Proponents argue that, by allowing traditional outdoor farms to revert to a natural state and reducing the energy costs needed to transport foods to consumers, vertical farms could significantly alleviate climate change produced by excess atmospheric carbon. Critics have noted that the costs of the additional energy needed for artificial lighting, heating and other vertical farming operations would outweigh the benefit of the building's close proximity to the areas of consumption.

One of the earliest drawings of a tall building that cultivates food for the purposes of consumption was published as early as Life Magazine 1909. The reproduced drawings feature vertically stacked homesteads set amidst a farming landscape. This proposal can be seen in Rem Koolhaas's Delirious New York. Koolhaas wrote that this 1909 theorem is

> *'The Skyscraper as Utopian device for the production of unlimited numbers of virgin sites on a metropolitan location' (1994, 82).*

Other architectural proposals that provide the seeds for the Vertical Farm project include Le Corbusier's Immeubles-Villas (1922) and SITE's Highrise of homes (1972). SITE's Highrise of homes, is a near revival of the 1909 Life Magazine Theorem. In fact, built examples of tower hydroponicums are quite well documented in the canonical text of "The Glass House" by John Hix. Images of the vertical farms at the School of Gardeners in Langenlois, Austria, and the glass tower at the Vienna International Horticulture Exhibition (1964) clearly show that vertical farms existed more than 40 years prior to contemporary discourse on the subject. Although architectural precedents remain valuable, the technological precedents that make vertical farming possible can be traced back to horticultural history through the development of greenhouse and hydroponic technology. Early building types or Hydroponicums were developed, integrating hydroponic technology into building systems. These horticultural building systems evolved from greenhouse technology, and paved the way for the modern concept of the vertical farm. The British Interplanetary Society developed a hydroponicum for lunar conditions and other building prototypes were developed during the early days of space exploration. During this era of expansion and experimentation, the first Tower Hydroponic Units were developed in Armenia.

The Armenian tower hydroponicums are the first built examples of a vertical farm, and is documented in Sholto Douglas' seminal text "Hydroponics: The Bengal System" first published in 1951 with data from the then-East Pakistan, today's Bangladesh, and the Indian state of West Bengal. Contemporary notions of vertical farming are predated by this early technology by more than 50 years. Contemporary precursors that have been published, or built, are Ken Yeang's Bioclimatic Skyscraper (Menara Mesiniaga, built 1992); MVRDV's PigCity, 2000; MVRDV's Meta City/ Datatown (1998–2000); Pich-Aguilera's Garden Towers (2001).

Ken Yeang is perhaps the most widely known architects that has promoted the idea of the 'mixed-use' Bioclimatic Skyscraper which combines living units and opportunities for food production.

Early prototypes of vertical farms, or "Tower Hydroponicums" existed in Armenia prior to 1951 during an era of hydroponic and horticultural building system research fueled by space exploration and a transatlantic technology race.

The latest version of these very idea is Dickson Despommier's "The Vertical Farm".

Dickson Despommier, a professor of environmental health sciences and microbiology at Columbia University in New York City, modernized the idea of vertical farming in 1999 with graduate students in a medical ecology class. Although much of Despommier's suggestions have been challenged and strongly criticized from an environmental science and engineering point of view, the idea's popularization in recent years has been largely the result of Despommier's assertion that food production can be transformed.

Despommier had originally challenged his class to feed the population of Manhattan (About 2,000,000 people) using 13 acres (5.3 ha) of usable rooftop gardens. The class calculated that, by using rooftop gardening methods, only 2 percent would be fed. Unsatisfied with the results, Despommier made an off-the-cuff suggestion of growing plants indoors, vertically. The idea sparked the students' interests and gained major momentum. By 2001 the first outline of a vertical farm was introduced and today scientists, architects, and investors worldwide are working together to make the concept of vertical farming a reality. In an interview with Miller-McCune.com, Despommier described how vertical farms would function:

Each floor will have its own watering and nutrient monitoring systems. There will be sensors for every single plant that tracks how much and what kinds of nutrients the plant has absorbed. You'll even have systems to monitor plant diseases by employing DNA chip technologies that detect the presence of plant pathogens by simply sampling the air and using snippets from various viral and bacterial infections. It's very easy to do.

Moreover, a gas chromatograph will tell us when to pick the plant by analyzing which flavenoids the produce contains. These flavonoids are what gives the food the flavors you're so fond of, particularly for more aromatic produce like tomatoes and peppers. These are all right-off-the-shelf technologies. The ability to construct a vertical farm exists now. We don't have to make anything new.

Architectural designs have been produced by Chris Jacobs and Andrew Kranis from Columbia University and Gordon Graff from the University of Waterloo's School of Architecture in Cambridge, ON. Together with Graff, and after disagreeing with Despommier's technical assumptions regarding energy and water balances in 2011, Tahbit Chowdhury and a multidisciplinary team from Waterloo's Dept. of Environmental Engineering and Dept. of Systems Design Engineering augmented the concepts with a focus on low-energy economically-intensive protein-production. Along with Chowdhury, others to have disagreed with Despommier's approach include Pierre Desrochers of the University of Toronto and Dennis T. Avery of the Centre for Global Food Issues, affiliated with the Hudson Institute.

Chowdhury and Graff applied advanced industrial engineering design philosophies to modernize current greenhouse technology as it pertains to hydroponics and aeroponics. The results of the Waterloo team's work showed that there is sufficient technical grounds to begin implementing Despommier's ideas for skyscrapers. However, Chowdhury and Graff showed that the designs will be dramatically different from what Despommier envisioned at Columbia.

Mass media attention began with an article written in *New York* magazine. Since 2007, articles have appeared in *The New York Times*, *U.S. News & World Report*, *Popular Science*, *Scientific American* and *Maxim*, among others, as well as radio and television features.

As of 2012, Vertical Harvest is working on raising funds for an urban, small-scale vertical farm in Jackson Hole, Wyoming.

Advantages

Several potential advantages of vertical farming have been discussed by Despommier. Many of these benefits are obtained from scaling up hydroponic or aeroponic growing methods.

Preparation for the Future

It is estimated that by the year 2050, close to 80% of the world's population will live in urban areas and the total population of the world will increase by 3 billion people. A very large amount of land may be required depending on the change in yield per hectare. Scientists are concerned that this large amount of required farmland will not be available and that severe damage to the earth will be caused by the added farmland. Vertical farms, if designed properly, may eliminate the need to create additional farmland and help create a cleaner environment.

Increased Crop Production

Unlike traditional farming in non-tropical areas, indoor farming can produce crops year-round. All-season farming multiplies the productivity of the farmed surface by a factor of 4 to 6 depending on the crop. With some crops, such as strawberries, the factor may be as high as 30.

Furthermore, as the crops would be sold in the same infrastructures in which they are grown, they will not need to be transported between production and sale, resulting in less spoilage, infestation, and energy required than conventional farming encounters. Research has shown that 30% of harvested crops are wasted due to spoilage and infestation, though this number is much lower in developed nations.

Despommier suggests that, if dwarf versions of certain crops are used (e.g. dwarf wheat developed by NASA, which is smaller in size but richer in nutrients), year-round crops, and "stacker" plant holders are accounted for, a 30-story building with a base of a building block (5 acres (20,000 m^2)) would yield a yearly crop analogous to that of 2,400 acres (9,700,000 m^2) of traditional farming.

Protection from Weather-related Problems

Crops grown in traditional outdoor farming suffer from the often suboptimal, and sometimes extreme, nature of geological and meteorological events such as undesirable temperatures or rainfall amounts, monsoons, hailstorms, tornadoes, flooding, wildfires, and severe droughts. The protection of crops from weather is increasingly important as global climate change occurs. "Three recent floods (in 1993, 2007 and 2008) cost the United States billions of dollars in lost crops, with even more devastating losses in topsoil. Changes in rain patterns and temperature could diminish India's agricultural output by 30 percent by the end of the century."

Because vertical farming provides a controlled environment, the productivity of vertical farms would be mostly independent of weather and protected from extreme weather events. Although the controlled environment of vertical farming negates most of these factors, earthquakes and tornadoes still pose threats to the proposed infrastructure, although this again depends on the location of the vertical farms.

Conservation of Resources

Each unit of area in a vertical farm could allow up to 20 units of area of outdoor farmland to return to its natural state, and recover farmlands due to development from original flat farmlands.

Vertical farming would reduce the need for new farmland due to overpopulation, thus saving many natural resources, currently threatened by deforestation or pollution. Deforestation and desertification caused by agricultural encroachment on natural biomes would be avoided. Because vertical farming lets crops be grown closer to consumers, it would substantially reduce the amount of fossil fuels currently used to transport and refrigerate farm produce. Producing food indoors reduces or eliminates conventional plowing, planting, and harvesting by farm machinery, also powered by fossil fuels. Burning less fossil fuel would reduce air pollution and the carbon dioxide emissions that cause climate change, as well as create healthier environments for humans and animals alike.

Organic Crops

The controlled growing environment reduces the need for pesticides, namely herbicides and fungicides. Advocates claim that producing organic crops in vertical farms is practical and the most likely production

Halting Mass Extinction

Withdrawing human activity from large areas of the Earth's land surface may be necessary to slow and eventually halt the current anthropogenic mass extinction of land animals.

Traditional agriculture is highly disruptive to wild animal populations that live in and around farmland and some argue it becomes unethical when there is a viable alternative. One study showed that wood mouse populations dropped from 25 per hectare to 5 per hectare after harvest, estimating 10 animals killed per hectare each year with conventional farming. In comparison, vertical farming would cause very little harm to wildlife.

Impact on Human Health

Traditional farming is a hazardous occupation with particular risks that often take their toll on the health of human laborers. Such risks include: exposure to infectious diseases such as malaria and schistosomes, exposure to toxic chemicals commonly used as pesticides and fungicides, confrontations with dangerous wildlife such as venomous snakes, and the severe injuries that can occur when using large industrial farming equipment. Whereas the traditional farming environment inevitably contains these risks (particularly in the farming practice known as "slash and burn"), vertical farming – because the environment is strictly controlled and predictable – reduces some of these dangers. Currently, the American food system makes fast, unhealthy food cheap while fresh produce is less available and more expensive, encouraging poor eating habits. These poor eating habits lead to health problems such as obesity, heart disease, and diabetes.

Urban Growth

Vertical farming, used in conjunction with other technologies and socioeconomic practices, could allow cities to expand while remaining largely self-sufficient food wise. This would allow for large urban centres that could grow without destroying considerably larger areas of forest to provide food for their people. Moreover, the industry of vertical farming will provide employment to these expanding urban centres. This may help displace the unemployment created by the dismantling of traditional farms, as more farm laborers move to cities in search of work. It is highly unlikely that traditional farms will become obsolete, as there are many crops that are not suited for vertical farming, and the production costs are currently much lower.

Energy Production

Vertical farms could exploit methane digesters to generate a small portion of its own electrical needs. Methane digesters could be built on site to transform the organic waste generated at the farm into biogas which is generally composed of 65% methane along with other gases. This biogas could then be burned to generate electricity for the greenhouse.

Technologies and Devices

Vertical farming relies on the use of various physical methods to become effective. Combining these technologies and devices in an integrated whole is necessary to make Vertical Farming a reality. Various methods are proposed and under research. The most common technologies suggested are:

- Greenhouse
- The Folkewall and other vertical growing architectures
- Aeroponics / Hydroponics / Aquaponics
- Composting
- Grow light
- Phytoremediation

- Skyscraper
- Controlled-environment agriculture

Plans

Despommier argues that the technology to construct vertical farms currently exists. He also states that the system can be profitable and effective, a claim evidenced by some preliminary research posted on the project's website. Developers and local governments in the following cities have expressed serious interest in establishing a vertical farm: Incheon (South Korea), Abu Dhabi (United Arab Emirates), and Dongtan (China), New York City, Portland, Ore., Los Angeles, Las Vegas, Seattle, Surrey, B.C., Toronto, Paris, Bangalore, Dubai, Shanghai and Beijing. The Illinois Institute of Technology is now crafting a detailed plan for Chicago. It is suggested that prototype versions of vertical farms should be created first, possibly at large universities interested in the research of vertical farms, in order to prevent failures such as the Biosphere 2 project in Oracle, Arizona.

In 2009, the world's first pilot production system was installed at Paignton Zoo Environmental Park in the United Kingdom. The project showcased a technological solution for vertical farming and provided a physical base to conduct research into sustainable urban food production. The produce is used to feed the zoos animals while the project enables evaluation of the systems and provides an educational resource to advocate for change in unsustainable land use practices that impact upon global biodiversity and ecosystem services,

In 2010, the Green Zionist Alliance proposed a resolution at the 36th World Zionist Congress calling on Keren Kayemet L'Yisrael (Jewish National Fund in Israel) to develop vertical farms in Israel.

In 2012, the world's first commercial vertical farm was opened in Singapore, developed by Sky Greens Farms, and is three stories high. They currently have over 100 towers that stand at nine metres tall.

Problems

Economics: Opponents question the potential profitability of vertical farming. A detailed cost analysis of start-up costs, operation costs, and revenue has not been done. The extra cost of lighting, heating, and powering the vertical farm may negate any of the cost benefits received by the decrease in transportation expenses. The economic and environmental benefits of vertical farming rest partly on the concept of minimizing food miles, the distance that food travels from farm to consumer. However, a recent analysis suggests that transportation is only a minor contributor to the economic and environmental costs of supplying food to urban populations. The author of the report, University of Toronto professor Pierre Desrochers, concluded that "food miles are, at best, a marketing fad."

Similarly, if the power needs of the vertical farm are met by fossil fuels, the environmental effect may be a net loss; even building low-carbon capacity to power the farms may not make as much sense as simply leaving the traditional farms in place, and burning less coal.

The initial building costs will be easily over $100 million, for a 60 hectare vertical farm. Office occupancy costs can be very high in major cities, with cities such as Tokyo, Moscow, Mumbai, Dubai, Milan, Zurich, and Sao Paulo ranging from $1850 to $880 per square metre, respectively.

Energy Use

During the growing season, the sun shines on a vertical surface at an extreme angle such that much less light is available to crops than when they are planted on flat land. Therefore, supplemental light, would be required in order to obtain economically viable yields. Bruce Bugbee, a crop physiologist at Utah State University, believes that the power demands of vertical farming will be too expensive and uncompetitive with traditional farms using only free natural light. The environmental writer George Monbiot calculated that the cost of providing enough supplementary light to grow the grain for a single loaf would be about $15. An article in the Economist argued that "even though crops growing in a glass skyscraper will get some natural sunlight during the day, it won't be enough" and "the cost of powering artificial lights will make indoor farming prohibitively expensive".

As "The Vertical Farm" proposes a controlled environment, heating and cooling costs will be at least as costly as any other tower. But there also remains the issue of complicated, if not more expensive, plumbing and elevator systems to distribute food and water throughout. Even throughout the northern continental United States, while heating with relatively cheap fossil fuels, the heating cost can be over $200,000/hectare.

To address this problem, The Plant in Chicago is building an anaerobic digester into the building. This will allow the farm to operate off the energy grid. Moreover, the anaerobic digester will be recycling waste from nearby businesses that would otherwise go into landfills.

Pollution

Regular greenhouse produce is known to create more greenhouse gases than field produce, largely due to higher energy use per kilogram of produce. With vertical farms requiring much greater energy per kilogram of produce, mainly through increased lighting, than regular greenhouses, the amount of pollution created will be much higher than that from field produce.

As plants acquire nearly all their carbon from the atmosphere, greenhouse growers commonly supplement CO_2 levels to 3-4 times the rate normally found in the atmosphere. This increase in CO_2, which has been shown to increase photosynthesis rates by 50%, contributes to the higher yields expected in vertical farming. It is not uncommon to find greenhouses burning fossil fuels purely for this purpose, as other CO_2 sources, like from furnaces, contain pollutants such as sulphur dioxide and ethylene which significantly damage plants. This means a vertical farm will require a CO_2 source, most likely from combustion, even if the rest of the farm is powered by 'green' energy. Also, through necessary ventilation, much CO_2 will be leaked into the city's atmosphere.

Greenhouse growers commonly exploit photoperiodism in plants to control whether the plants are in a vegetative or reproductive stage. As part of this control, growers will have the lights on past sunset and before sunrise or periodically throughout the night. Single story greenhouses are already a nuisance to neighbours because of light pollution, a 30 story vertical farm in a densely populated area will surely face problems because of its light pollution.

Hydroponics greenhouses regularly change the water, meaning there is a large quantity of water containing fertilizers and pesticides that must be disposed of. While solutions are currently being worked on, the most common method of simply spreading the mixture over a sufficient area of neighbouring farmland or wetlands would be more difficult for an urban vertical farm.

3

Medicinal Plants

Medicinal plants have been identified and used throughout human history. Toxic plants even have use in pharmaceutical development. Angiosperms (flowering plants) were the original source of most plant medicines. Some herbs and spices come from flowering plants.

Plants have the ability to synthesize a wide variety of chemical compounds that are used to perform important biological functions, and to defend against attack from predators such as insects, fungi and herbivorous mammals. Many of these phytochemicals have beneficial effects on long-term health when consumed by humans, and can be used to effectively treat human diseases. At least 12,000 such compounds have been isolated so far; a number estimated to be less than 10% of the total. These phytochemicals are divided into (1) primary metabolites such as sugars and fats, which are found in all plants; and (2) secondary metabolites – compounds which are found in a smaller range of plants, serving a more specific function. For example, some secondary metabolites are toxins used to deter predation and others are pheromones used to attract insects for pollination. It is these secondary metabolites and pigments that can have therapeutic actions in humans and which can be refined to produce drugs—examples are inulin from the roots of dahlias, quinine from the cinchona, morphine and codeine from the poppy, and digoxin from the foxglove. Chemical compounds in plants mediate their effects on the human body through processes identical to those already well understood for the chemical compounds in conventional drugs; thus herbal medicines do not differ greatly from conventional drugs in terms of how they work. This enables herbal medicines to be as effective as conventional medicines, but also gives them the same potential to cause harmful side effects.

Most cultures have a tradition of using plants medicinally. In Europe, apothecaries stocked herbal ingredients for their medicines. In the Latin names for plants created by Linnaeus, the word *officinalis* indicates that a plant was used in this way. For example, the marsh mallow has the classification *Althaea officinalis*, as it was traditionally used as an emollient to soothe ulcers. Ayurvedic medicine, herbal medicine and traditional Chinese medicine are other examples of medical practices that incorporate medical uses of plants. Pharmacognosy is the branch of modern medicine about medicines from plant sources. Plants included here are those that have been or are being used medicinally, in at least one such medicinal tradition.

Modern medicine now tends to use the active ingredients of plants rather than the whole plants. The phytochemicals may be synthesized, compounded or otherwise transformed to make pharmaceuticals. Examples of such derivatives include Digoxin, from digitalis; capsaicine, from chili; and aspirin, which is chemically related to the salicylic acid found in white willow. The opium poppy continues to be a major industrial source of opiates, including morphine. Few traditional remedies, however, have translated into modern drugs, although there is continuing research into the efficacy and possible adaptation of traditional herbal treatments.

Safed Musli

Chlorophytum borivilianum is a herb with lanceolate leaves, from tropical wet forests in peninsular India. The Hindi name is safed musli.

It is cultivated and eaten as a leaf vegetable in some parts of India, and its roots are used medicinally as a sex tonic under the name *safed musli*. The medicinal value is thought to derive from its saponin content, up to 17 percent by dry weight. It has also recently been suggested that it may produce an aphrodisiac agent. As medicinal demand has increased, the plant has been brought under cultivation. The saponins and alkaloids present in the plant are the source of its alleged aphrodisiac properties.

This herb is used in laddoos for women postpartum.

In traditional Indian medicine it is used as 'Rasayan' or adaptogen.

Curculigo orchioides (Gaertn) is a small perennial herb with elongated tuberous root stock having several lateral roots with rosette of sessile or short petiolate, linear lanceolated membranous leaves close to the ground level.

Leaves: Sessile or short petiolate, with sheathing bases, 15 to 45 cms. by 1.5 2.5 cms., liner or linear lanceolate, membranous, glabrous or sparsely softly hairy. The leaf tip in contact with soil produces adventitious roots

Inflorescence: Axillary, scapose, racemose, scape short, clavate flattened with the pedicels, bracts and overy concealed in the leaf sheaths. The lower flowers on the scape are mostly bisexual and the upper staminate, flowers: epigynous: bright yellow, bisexual or unisexual with lanceolate membraneous bracts. Perianth:

Gamophyllous, rotate, six partite, perianth segments 12 to 16 mm. long, elliptic belong acute, hairy on the back stamens:-6, tilarnents very small, adnate to the base of the perianth lobes, anthers linear erect. Ovary: inferior, tricarpellary, syncarpous with a fairly long, slender beak and stipe, ovules 6-8 with long finical styles 1 short, stigma 3 cleft with the lobes oblong, erect and apprised.

Fruit: Capsule, 4-seeded with a slender beak and spongy septa, seeds oblong, black and shiny with crustaceans testa. Root stock stout, short or elongate, more or less cylindrical, 4 to 15 cm. long, average being 7.4 cm., 0.5- 1. cm. wide. The upper portion of the root stock remains clothed with the withered leaf bases and with copious lateral roots which are long, almost whitish or yellowish gray, shrieked. Externally the root stock is yellowish brown, thin. Fracture is starchy, odour indistinct; taste slightly bitter and mucilaginous.

Tiwarri and Misra (1976) have identified a new glycoside 5,7-dimethoxy myricetin 3-0 &-L-xylopyranosyl 4-0 3 -D glucopyranoside from the rhizome of Curculigo orchioides. The tubers of the plant contain free sugars, 7.56%: mucilage, 8.12%: hemicelluloses, 12 to 15% and other polysachharides 17.01 %.

Cultivation

In order to the meet market demand for tubers of the Safed Musli private growers have started cultivating Chlorophytum borivilianum species, which is a fast growing species. The technical details and the cultivation methodology arc as under.

1. Name of the crop in Hindi: Safed Musli.
2. Botanical Name - Chlorophytum borivilianum
3. Application Tonic.
4. Growth of roots-Tuberous (usable part is the roots of the plants).
5. Flowering 4 P(3+3), A3i G White in colour.

6. Chromosomes - 16
7. Seeds — Black triangular two seeds arc present in each locule.
8. Family - Liliaceae.
9. Proposed method of Planting of the crop:

   A: pure cropping in circularly raised beds (i2." height)

   B: lntercropping (around trees).
10. Crop maturity- 10 to 11 months duration.

The Preparation of Land

The land is required to be prepared in the month of April-May. After deep ploughing of the fields, 20 to 50 trolleys (40 to 80 tonnes) per acre of FYM is mixed in the soil. If the soil is of clay nature them soil conditioner at the rate of one tonne per acre could be used to condition it. Green manure can also be used to enrich the soil. For this purpose, the seeds of Sunhemp (boru) @ 60 kg. per acre could be sown in the field (proposed for the cultivation of Musli) during March- April and before they start flowering they are cut and mixed into the soil. Sometimes the field is irrigated also for the easy decomposition of green manure. If the soil gets clods, then 3-4 tillers are given for better pulverisation of soil. After preparation of land, beds are prepared. For the proper growth of the tubers of Mush and to facilitate the proper drainage, raised beds are prepared. Although the size of these beds could be as per facility and ease but generally beds with height up to one foot and a width of 3.5 ft. is recommended for the proper growth and development of Mush tubers.

Sowing of Musli

Being a Kharif crop, the sowing of mush starts with the first shower of the monsoons. Although its seeds could also be used for sowing, but for better result, the tubers suit most. Thus form the old bunch of tubers their fingers are separated. It is ensured that some part of that crown/disc remains in intact with all the fingers which are to be used for sowing. Generally these fingers are planted at a distance of 10 and a. total 32000 fingers are required per acre with an approximate weight of 400 to 500 kgs.

It is advisable to treat the fingers or tubers planting material either with urine of cow (proportion of cow urine and water should be 1:10) or with Bovestin and Streptocylin (30gm of Bovestin and 3 gins of Streptocylin to be mixed in 15 litres of water), before planting in order to keep it free from fungus.

Controlling of Weeds

Generally two weeding of the crop are required to free it from the weeds. The first within 20-25 days of sowing and the other 20-25 days later. The fields must be kept free from weeds to obtain the proper yield for the crop.

Fertilizers and Pesticides Required for the Crop

Apart from the FYM used during the preparation of fields, it is advisable to apply NPK (5 0:40:40 kg. per acre) in the field after the first rain fall. The crop is generally free from most of the pests and diseases. Sometimes water logging causes damage to tuber roots which can be cured by stopping further irrigation. For the control of illi and other pests, one has to spray thyrum 5 ml. litre every month. This could be applied as a precautionary measure also. The fungus that attacks Safed Musli "Fusarium". Trichoderma virdi is used to destroy the attack of this fungus.

Harvesting of the Crop

After 3-3.5 months of sowing (generally in the month of Oct/Nov the leaves of Musli start yellowing. Subsequently they become dry and fall off and get detached from the tuber/disc. Some research scientists are of the opinion that this is maturity stage of the crop. When the crop attains this stage then it means the crop cycle is complete, hence the tubers should be dug out. But people engaged in the cultivation of musli in the most professional manner do not agree with it. They find that even after the drying up of the leaves, tubers/disc fingers of mush keep on maturing in the soil. So if at this stage the tubers are dug out they will not be fully matured and hence will lack some medicinal properties thus, these tubers should be left in the soil for sometime where they keep on maturing.

During this time also the moisture level in the soil should be maintained. By Jan-Feb the skin of tubers mature and it turns to dark brown. This is the right time to dig out the tubers thus on attaining this stage, the tubers should be dug out.

After digging out the tubers they are properly washed. Of these tubers, the longer and healthy fingers are. detached from the tubers, whereas the smaller ones are kept to be used as planting material for the next season. The longer and healthy fingers already detached from the tubers are then taken for processing. During processing the skin of the fingers are peeled out with a stainless steel knife and then they are kept in the sun for drying. Within 3 days the fingers dry up. These are then packed in the polythene bags and sent to the market.

Production Per Acre

On an average this crop gives an yield of 20-30 quintals of wet mush per acre. After peeling and drying nearly 20% (4-4.5 quintals) dry mush is finally obtained. In the indigenous market, the present rate of dry mush ranges between Rs. 800-1800 per kg., whereas it fetches beyond Rs. 3000 per kg. in the international market.

Uses of Safed Musli

Safed Musli has very good ayurvedic medicinal use. It is a rich source of over 25 alkaloids, vitamins, minerals, proteins, carbohydrates, steroids saponins and polysachharides etc. Musli or Talamuli ha.s been used iii the indigenous systeni of medicine for a long period. According to Bhavprakash the drug is sweet, cooling, mucilaginous increases Kapha and reduces 'Pitta daha and acts as stimulant, it gives strength Mush prepared as a paste with goats milk or honey and applied locally over the face, brightens the complexion of the face.

It is extensively used by the Ayurvedic practitioners for a wide variety of ailments and particularly an ingredient of aphrodisiac preparations.

In Raj Nighantu it has been described that the Kali mush is more effective than Sweta mush, which is botanically identified as Chlorophytum tuberosum. The root stock extracts is effective mainly on the urinary system and is considered to be diuretic in action. it is given in dysuria, poluria, gonorrhoea in addition to the menorrhagia, lcueorrhoea, piles and leucoderma. The drug is considered to be alternative appetiser, fattening and restorative presently a number of health tonics (sexual tonics) are prepared from it.

In Unani system of medicine, the drug is employed in bronchitis, opthalmic conditions, vomiting, than hoca, dyspepsia, lumbago, pain in the joints and in the diseases of nerves. It is considered to have carminative, tonic, aphrodisiac and anti-pyretic properties.

Marketing

The tubers of the Safed Musli have got very good Ayurvedic medicinal properties. It is a rich source of over 25 alkaloids.

The outer layer of the tubers is removed and it is dried in the Sunlight. The dried tubers are then bagged and weighed. The tubers in such dried form are then sent to major markets like Delhi, Mumbai etc. The Ayurvedic medicine companies directly contact the cultivators who can sale the tubers directly to them as per their requirement. Those who want to grow mush on their farm they to have to buy fresh tubers from the other cultivators. For storing fresh stock cold storage is required.

Legal Aspects

Safed Musli is a rare herb and hence its preservation and conservation carries lot of importance. As per the provisions embodied in the Indian Forest Act 1927 Safed Musli and its derived products come under the purview of definition of forest produce. Hence its transit need be covered by a transit pass. This will ensure to check illegal transportation of the herb. The produce of the cultivators can be allowed to be transported after they have procured a legal procurement certificate from forest authorities. The existing laws can be amended so as to check the illegal transportation of Safed Musli.

Ashwagandha

Withania somnifera, known commonly as *ashwagandha*, Indian ginseng, poison gooseberry, or winter cherry, is a plant in the Solanaceae or nightshade family. Several other species in the genus *Withania* are morphologically similar. It is used as an herb in Ayurvedic medicine. This species is a short shrub growing 35 to 75 centimetres tall. Tomentose branches extend radially from a central stem. The flowers are small and green. The ripe fruit is orange-red.

Cultivation

Withania somnifera is cultivated in many of the drier regions of India, such as Mandsaur District of Madhya Pradesh, Punjab, Sindh, Gujarat,and Rajasthan. It is also found in Nepal. In the wild, Ashwagandha (Withania Somnifera) grows profusely in most areas of South Asia and many closely related Withania species occur as far away as Northern Africa. As a commercial crop Ashwagandha cultivation is carried out mostly in Utter Pradesh and Madhya Pradesh in India. It is generally grown in fields characterized by slightly basic soils that have good drainage. Light red soils are suitable for this herb. The herb grows best in areas receiving about 500-750mm rainfall and at altitudes ranging from 600 to 1200 metres above sea level. The optimal temperature range is 20-32 degrees C. Due to the hardiness of the plant, Ashwagandha has historically been grown in areas that are not well irrigated and therefore not suitable for food crops. It is a cash crop in fields that would not be suitable to produce more profitable plants. Ashwagandha cultivation is possible even in mild drought conditions.

Germination from Seeds

The Ashwagandha seeds are first planted in small areas or in a nursery in lines to allow them to germinate. The seeds are sown about

2 cm deep into the soil at a distance of about 10 cm. The amount of seeds germinated should be based upon a final transplanted density of about 1kg of Ashwagandha seeds per hectare. The line method of sowing is preferred at this germination stage as it promotes the development of a health root system compared to being sown using the broad casting method. The germination is best carried out just before the arrival of the monsoon season; and a few light showers during this stage produce the optimal seedlings. Once the young plants are about 30 days old, they are ready for transplanting into the main farming fields. In these fields they are planted at a distance of approximately 60 cm into well ploughed and aerated soil. While the 30 day age of the seedlings is a guideline, in practice the transplanting correspondence to the actual arrival of the first monsoon rains.

Watering and Fertilization

Given the ultimate medicinal use of the herb, the plants are usually not fertilized. In some cases organic fertilizer is used. The fields are also usually left to natural irrigation and artificial watering is only done in severe drought conditions where the survival of the crop is in jeopardy. It is this very hardy nature of Ashwagandha cultivation that has made this sort of farming popular on otherwise unproductive land.

Harvesting

The new crop will be ready for harvesting in about 180 days. This usually results in a harvest around early January. The plants will start flowering from mid December onwards and it is determined if they are ready to harvest by observing the development of the red berry like fruits. The whole plant along with the roots is removed from the soil. The roots and berries are the main parts used. The plant is separated from the root about 2cm above where the aerial part begins. The berries are hand picked and separated from the plant. Both the roots and berries ate then subject to drying. The roots will most likely be cut down into smaller pieces while the berries are dried whole. Once fully dry, the roots are transferred for further processing while the berries are crushed to separate the seeds. A typical harvest is usually 350 to 400kg of dried Ashwagandha root per 1 hectare of land.

Pathology

Withania somnifera is prone to several pests and diseases. Leaf spot disease caused by *Alternaria alternata* is the most prevalent disease, which is most severe in the plains of Punjab, Haryana, and Himachal Pradesh. Biodeterioration of its pharmaceutically active components during leaf spot disease has been reported. The *Choanephora cucurbitarum* causes a stem and leaf rot of *Withania somnifera Oxyrachis tarandus*, a treehopper, feeds on the apical portions of the stem, making them rough and woody in appearance and brown in colour. The apical leaves are shed and the plant gradually dies. The carmine red spider mite (*Tetranychus urticae*) is the most prevalent pest of the plant in India.

Culinary use

The berries can be used as a substitute for rennet in cheesemaking.

Biochemistry

The main chemical constituents are alkaloids and steroidal lactones. These include tropine and cuscohygrine. The leaves contain the steroidal lactones, withanolides, notably withaferin A, which was the first to be isolated from the plant.

Traditional Medicinal Uses

The plant's long, brown, tuberous roots are used for medicinal purposes.

In Ayurveda, the berries and leaves are applied externally to tumors, tubercular glands, carbuncles, and ulcers. The roots are used to prepare the herbal remedy *ashwagandha*, which has been traditionally used to treat various symptoms and conditions.

In two published clinical trials of *W. somnifera*, the side effects were not significantly different from those experienced by placebo-treated individuals. *Ashwagandha* exhibits greater clinical benefit than psychotherapy in mental health (anxiety level), concentration, fatigue, social functioning, vitality, and overall quality of life.

Multiple studies have shown positive effects in reducing physiological and psychological stress as well as restoring and increasing energy levels. More than one study had findings showing a significant improvement in both cardiovascular and respiratory endurance.

Amla

Phyllanthus emblica (syn. *Emblica officinalis*), the Indian gooseberry, or Dhatrik (in Maithili), or amla from Sanskrit amalika.

The tree is small to medium in size, reaching 8 to 18 m in height, with a crooked trunk and spreading branches. The branchlets are glabrous or finely pubescent, 10–20 cm long, usually deciduous; the leaves are simple, subsessile and closely set along branchlets, light green, resembling pinnate leaves. The flowers are greenish-yellow. The fruit is nearly spherical, light greenish yellow, quite smooth and hard on appearance, with six vertical stripes or furrows.

Ripening in autumn, the berries are harvested by hand after climbing to upper branches bearing the fruits. The taste of Indian gooseberry is sour, bitter and astringent, and it is quite fibrous. In India, it is common to eat gooseberries steeped in salt water and turmeric to make the sour fruits palatable. It is also used to straighten hair.

Varieties

The most popular cultivable varieties of amla are Banarasi, NA 7, Krishna, Kanchan, Chakaiya, BSR 1.

Soil and Climate

Amla is a subtropical plant and prefers dry climate. It is a hardy plant and can be grown in variable soil conditions. The crop can tolerate salinity and alkalinity.

Planting Material

Seedlings, Grafts and Buddings are used for planting.

Planting

Planting is done during July-August with a spacing of 6 x 6 m in pits of 1x1m or 1.25 x 1.25m.

Irrigation

Irrigate the plants initially for establishment. No irrigation is required during rainy and winter season. Drip irrigation is appropriate which can save water upto 40-45%.

Manures and Fertilizers (per plant/year)

Manures and Fertilizers	*Bearing tree*
FYM	10 kg
N	200 g
P	500 g
K	200 g

Manuring should be done immediately after pruning.

Training and Pruning

The main branches should be allowed to appear at a height of 0.75-1 m above the ground level. Plants should be trained to modified central leader system. Two to four branches with wide crotch angle, appearing in the opposite directions should be encouraged in early years.

During March – April, prune and thin the crowded branches to provide maximum fruit bearing area in the tree.

Stem Swelling

Control

- Stem swelling / bulging can be controlled by proper pruning
- Prophylactic spray of Monocrotophos 2.0 ml/lit can be sprayed twice at fortnight intervals.

Plant protection

Pest

Gall caterpillar: Young caterpillars bore into the apical portion of the shoot during rainy season and make tunnel. Due to this, apical regrowth is checked, side shoots develop below the gall and subsequent growth in following season is greatly hampered. Cutting off the infected apices and prophylactic spray of systemic insecticide like Dimethoate 0.03 percent will control the pest.

Bark eating cater pillar: Damages the stem and branches of grown up trees by eating the bark. The affected portion should be cleared and few drops of kerosene should be applied in holes to keep this in control.

Disease

Rust: Rust appears as circular reddish solitary or gregarious on leaves and also on fruits. Spray 0.2 per cent Mancozeb at an interval of 7 to 28 days during July to September.

Yield: The crop yields about 100 kg/tree annually.

Medical Research

Indian gooseberry has undergone preliminary research, demonstrating *in vitro* antiviral and antimicrobial properties. There is preliminary evidence *in vitro* that its extracts induce apoptosis and modify gene expression in osteoclasts involved in rheumatoid arthritis and osteoporosis. It may prove to have potential activity against some cancers. One recent animal study found treatment with *E. ofûcinalis* reduced severity of acute pancreatitis (induced by L-arginine in rats). It also promoted the spontaneous repair and regeneration process of the pancreas occurring after an acute attack.

Experimental preparations of leaves, bark or fruit have shown potential efficacy against laboratory models of disease, such as for inflammation, cancer, age-related renal disease, and diabetes.

A human pilot study demonstrated a reduction of blood cholesterol levels in both normal and hypercholesterolemic men with treatment. Another recent study with alloxan-induced diabetic rats given an aqueous amla fruit extract has shown significant decrease of the blood glucose, as well as triglyceridemic levels and an improvement of the liver function caused by a normalization of the liver-specific enzyme alanine transaminase activity.

Chemical Research

Although these fruits are reputed to contain high amounts of ascorbic acid (vitamin C), 445 mg/100g, the specific contents are disputed, and the overall antioxidant strength of amla may derive instead from its high density of ellagitannins such as emblicanin A (37%), emblicanin B (33%), punigluconin (12%) and pedunculagin (14%). It also contains punicafolin and phyllanemblinin A, phyllanemblin other polyphenols: flavonoids, kaempferol, ellagic acid and gallic acid.

Cultural and Religious Significance

In the Sanskrit Buddhist tradition half an amalaka fruit was the final gift to the Buddhist sangha by the great Indian emperor Asoka. This is illustrated in the Asokavadana in the following verses: "A great donor, the lord of men, the eminent Maurya Asoka, has gone from being lord of Jambudvipa [India] to being lord of half a myrobalan." (Strong, 1983, p. 99) This deed became so famous that a stupa was created to mark the place of the event in modern day Patna and was known as the Amalaka stupa.

According to Hindu tradition, Adi Shankara of Kerala composed and recited the Kanakadhara stotram in praise of Mahalakshmi to make a poor Brahmin lady get wealth, in return for a single amla presented to him as bhiksha on an auspicious dwadashi day.

According to a Tamil legend, Avvaiyar, a female poet, ethicist and political activist of the Sangam period was gifted with one amla by King Athiyaman to give her long life.

The tree is considered sacred by Hindus as the Vishnu is believed to dwell here. The tree is worshipped on Amalaka Ekadashi.

In other Hindu myths, Aamla is said to have originated from the drops of Amrit which spilled on earth accidentally, due to the fight of Gods and Demons after ksheera sagar manthan. And hence also this religious belief makes claims that it almost cures every disease and is also good in extending the longevity of life.

Traditional Uses of Amlaki

Medicinal use: In traditional Indian medicine, dried and fresh fruits of the plant are used. All parts of the plant are used in various Ayurvedic/Unani medicine (*Jawarish amla*) herbal preparations, including the fruit, seed, leaves, root, bark and flowers. According to Ayurveda, aamla fruit is sour (*amla*) and astringent (*kashaya*) in taste (*rasa*), with sweet (*madhura*), bitter (*tikta*) and pungent (*katu*) secondary tastes (*anurasas*). Its qualities (*gunas*) are light (*laghu*) and dry (*ruksha*), the postdigestive effect (*vipaka*) is sweet (*madhura*), and its energy (*virya*) is cooling (*shita*).

According to Ayurveda, aamla balances all three doshas. While aamla is unusual in that it contains five out of the six tastes recognized by Ayurveda, it is most important to recognize the effects of the "virya", or potency, and "vipaka", or post-digestive effect. Considered in this light, aamla is particularly helpful in reducing *pitta* due to its cooling energy. It also balances both Pitta and *vata* by virtue of its sweet taste. The *kapha* is balanced primarily due to its drying action. It may be used as a *rasayana* (rejuvenative) to promote longevity, and traditionally to enhance digestion (*dipanapachana*), treat constipation (*anuloma*), reduce fever (*jvaraghna*), purify the blood (*raktaprasadana*), reduce cough (*kasahara*), alleviate asthma (*svasahara*), strengthen the heart (*hrdaya*), benefit the eyes (*chakshushya*), stimulate hair growth (*romasanjana*), enliven the body (*jivaniya*), and enhance intellect (*medhya*).

In Ayurvedic polyherbal formulations, Indian gooseberry is a common constituent, and most notably is the primary ingredient in an ancient herbal *rasayana* called *Chyawanprash*. This formula, which contains 43 herbal ingredients as well as clarified butter, sesame oil, sugar cane juice, and honey, was first mentioned in the Charaka Samhita as a premier rejuvenative compound.

In Chinese traditional therapy, this fruit is called *yuganzi*, which is used to cure throat inflammation.

Emblica officinalis tea may ameliorate diabetic neuropathy due to aldose reductase inhibition. In rats it significantly reduced blood glucose, food intake, water intake and urine output in diabetic rats compared with the non diabetic control group.

Culinary Use

Particularly in South India, the fruit is pickled with salt, oil, and spices. Aamla is eaten raw or cooked into various dishes. In Andhra Pradesh, tender varieties are used to prepare *dal* (a lentil preparation), and *amle ka murabbah*, a sweet dish indigenous to the northern part of India (wherein the berries are soaked in sugar syrup for a long time till they are imparted the sweet flavour); it is traditionally consumed after meals.

Other Uses

Popularly used in inks, shampoos and hair oils, the high tannin content of Indian gooseberry fruit serves as a mordant for fixing dyes in fabrics. Amla shampoos and hair oil are traditionally believed to nourish the hair and scalp and prevent premature grey hair.

Gingseng

Ginseng is any one of 11 species of slow-growing perennial plants with fleshy roots, belonging to the genus *Panax* of the family Araliaceae.

Ginseng is found in North America and in eastern Asia (mostly Korea, northeast China, Bhutan, eastern Siberia) and in New Zealand, typically in cooler climates. *Panax vietnamensis*, discovered in Vietnam, is the southernmost ginseng known. This article focuses on the series *Panax* ginsengs, which are the adaptogenic herbs, principally *Panax ginseng* and *P. quinquefolius*. Ginseng is characterized by the presence of ginsenosides. Siberian ginseng (*Eleutherococcus senticosus*) is in the same family, but not genus, as true ginseng. Like ginseng, it is considered to be an adaptogenic herb. The active compounds in Siberian ginseng are eleutherosides, not ginsenosides. Instead of a fleshy root, Siberian ginseng has a woody root.

Economics

In 2010, nearly all of the world's 80,000 tons of ginseng in international commerce was produced in four countries: South Korea, China, Canada, and the United States. The product was marketed in over 35 countries. Sales exceeded $2.1 billion, of which half came from South Korea. Historically, Korea has been the largest provider, and China the largest consumer. Control over the ginseng fields was an issue in the 16th century.

Traditional Uses

The root is most often available in dried form, either whole or sliced. Ginseng leaf, although not as highly prized, is sometimes also used; as with the root, it is most often available in dried form. Folk medicine attributes various benefits to oral use of American ginseng and Asian ginseng (*P. ginseng*) roots, including roles as an aphrodisiac, stimulant, type II diabetes treatment, or cure for sexual dysfunction in men.

Ginseng may be included in small doses in energy drinks or herbal teas, such as ginseng coffee. It may be found in hair tonics and cosmetic preparations, as well, but those uses have not been shown to be clinically effective.

Research

Ginsenosides, unique compounds of the *Panax* species, are under basic and clinical research to investigate their potential for use in medicine.

Much research has been done on Ginseng with varying results. A study showed taking Siberian Ginseng daily can increase the number of white blood cells including activity of the T cells as well as the cytotoxic T cells and natural killer cells that eliminate invading cells and those that have been virally infected. A double blind-study with 93 volunteers with herpes simplex virus 2 (which can lead to genital herpes) found that there was a reduction (of 50%) in the number of outbreaks. Those that did occur were less severe and did not last as long. Ginseng is known to contain phytoestrogens.

Common Classification

Asian ginseng (root): *Panax ginseng* is available commercially as fresh, red, and white ginsengs; wild ginseng is used where available.

Red Ginseng

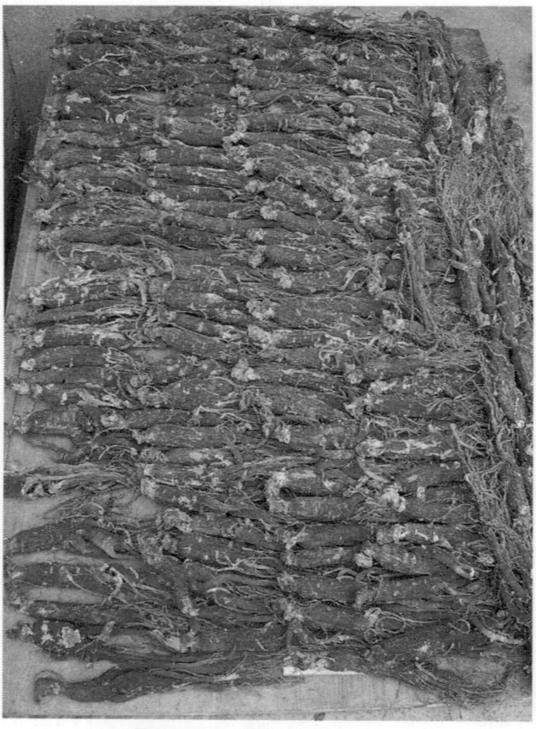

Figure: *Red ginseng.*

Red ginseng has been peeled, heated through steaming at standard boiling temperatures of 100 °C (212 °F), and then dried or sun-dried. It is frequently marinated in an herbal brew which results in the root becoming extremely brittle. It is more common as herbal medicine than white ginseng.

Fresh Ginseng

Fresh ginseng is the raw product. Its use is limited by availability.

White Ginseng

White ginseng, native to America, is fresh ginseng which has been dried without being heated. It is peeled and dried to reduce the water content to 12% or less. White ginseng air-dried in the sun may contain less of the therapeutic constituents. It is thought by some that enzymes contained in the root break down these constituents in the process of drying. Drying in the sun bleaches the root to a yellowish-white colour.

Wild Ginseng

Wild ginseng grows naturally and is harvested from wherever it is found. It is relatively rare, and even increasingly endangered, due in large part to high demand for the product in recent years, which has led to the wild plants being sought out and harvested faster than new ones can grow (it requires years for a root to reach maturity). Wild ginseng can be either Asian or American, and can be processed to be red ginseng.

Woods-grown American ginseng programmes in Vermont, Maine, Tennessee, Virginia, North Carolina, West Virginia and Kentucky, and United Plant Savers have been encouraging the planting of ginseng both to restore natural habitats and to remove pressure from any remaining wild ginseng, and they offer both advice and sources of rootlets. Woods-grown plants have a value comparable to wild-grown ginseng of similar age.

Partially germinated ginseng seeds harvested the previous Fall can be planted from early Spring until late Fall, and will sprout the following Spring. If planted in a wild setting and left to their own devices, they will develop into mature plants which cannot be distinguished from native wild plants. Both Asian and American partially germinated ginseng seeds can be bought from May through December on various eBay sales. Some seed sales come with planting and growing instructions.

P. Quinquefolius *American Ginseng (root)*

According to traditional Chinese medicine, American ginseng promotes yin energy, cleans excess yang and calms the body. The reason it has been claimed that American ginseng promotes yin (shadow, cold, negative, female) while Asian ginseng promotes yang (sunshine, hot, positive, male) is that, according to traditional Chinese medicine, things living in cold places or northern side of mountains or southern side of rivers are strong in yang and vice versa, so the two are balanced. Chinese/Korean ginseng grows in Manchuria and Korea, the coldest area known to many Koreans in ancient times. Thus, ginseng from there is supposed to be very *yang*.

Originally, American ginseng was imported into China via subtropical Guangzhou, the seaport next to Hong Kong, so Chinese doctors believed American ginseng must be good for *yang*, because it came from a hot area. They did not know, however, that American ginseng can only grow in temperate regions. Nonetheless, the root is legitimately classified as more *yin* because it generates fluids. Most North American ginseng is produced in the Canadian provinces of Ontario and British Columbia and the American state of Wisconsin, according to Agri-food Canada. *P. quinquefolius* is now also grown in northern China. The aromatic root resembles a small parsnip that forks as it matures. The plant grows 6 to 18 in tall, usually bearing three leaves, each with three to five leaflets two to five inches long.

Other Plants Sometimes called Ginseng

These mostly "adaptogenic" plants are sometimes referred to as ginsengs, but they are either from a different family or genus. Only *jiaogulan* actually contains compounds closely related to ginsenosides, although ginsenosides alone do not determine the effectiveness of ginseng. Since each of these plants has different uses, one should research their properties before using.

Cultivation

Ginseng is sometimes a difficult plant to grow. It is a plant that requires commitment for the fact that roots are best when left to maturity for 5-10 years. It is a committed gardener that decides to plant a crop, waiting 5-10 years to harvest the best and most potent roots is a task filled with risk, patience, and an eye to the distant future.

Long used as an herbal remedy, Ginseng Panax brings the mystique of the Orient to your patio, backyard, perennial garden or woodland plot. Stunning in bloom, this conversation piece is particularly attractive when massed beneath a grove of shade trees, or cascading down a forested hillside towards a rocky stream.

Ginseng did not get its reputation as a tonic and healing herb for nothing. The quest to grow wild simulated ginseng can be more than a hobby. It can be a reconnection with having faith in Nature's ability to provide cures in things that are healthy for us. Growing in the deep woods, it is not naturally a plant of the cultivated field. To grow it in its natural habitat is to remain connected to the forest, for the trees are your allies.

You can recreate the shady environment and conditions to grow ginseng in a flower pot, or locate the perfect spot in your yard or garden. If you have nice woods on your property you might consider a crop.

It is best to transplant the rootlets in spring and fall. Usually March and April just before they begin to bud. In the fall after the berries have fallen off is also a good time. Handle root delicately, taking care not to disturb the root during all transplanting procedures. Plant it in the fall and leave it alone until it comes up in the spring.

Ginseng requires 80 percent shade to thrive and prefers a deep rich loamy soil. Organic humus works well. It also requires good drainage.

The Wild-Simulated Method

A Method called wild-simulated can be used to grow ginseng without fungicide sprays and expensive establishment costs. The prices paid for genuine wild-simulated roots approach that of wild roots. It takes an expert to really tell them apart.

Ginseng cultivation is risky. The crop can be stolen or eaten by rodents. The chosen site may not be right. Plant diseases will certainly occur. On the other hand if the right conditions can be found, wild-simulated cultivation can provide supplemental income and expose us to a form of agriculture that is in harmony with the forest. It is a completely natural process.

The first step is site selection. Favourable soil and moisture conditions are most often associated with north or east facing slopes with at least 75 percent shade canopy. This is dense shade we're talking about. The best shade is provided by deep rooted, deciduous trees such as Poplar and Oaks. Ginseng grows best in a moist well drained soil. That is a contradiction, but these soils do exist. Successful growth of ginseng most often occurs in sites where herbaceous woodland plants such as Jack-in-the-Pulpit, Bloodroot, Solomon's Seal and ferns are thriving. If no herbaceous plants are growing on the forest floor, ginseng will probably not grow there. Excellent soil drainage is essential and swampy or clay soil must be avoided.

In the wild-simulated method, stratified ginseng seed is planted in the fall when the trees lose their leaves. In some locations, clearing of undergrowth will be necessary. If the site is sufficiently shaded, there should not be a great deal of competitive weed growth. Avoid dense patches of weeds, and disturb the site as little as possible to reduce the spread of fungal disease.

It is recommended that only stratified seed be planted. Freshly harvested seed that is placed in the sand is known as 'green seed'. Planting stratified seed means that the seed is only vulnerable for a very short time period (as opposed to 18 months in natural conditions) and survival rates are much greater.

The only tools needed to plant wild simulated ginseng are a rake and a garden hoe. It is a good idea to plants seeds in generally defined beds that are five feet wide and up to fifty feet long. The beds can be separated by three foot walkways. The beds should run up and down the slope rather than across the slope for better air and water drainage around the plants. Rake the leaves on the forest floor away from the bed right down to the topsoil. Using the hoe, make three narrow furrows 18 inches apart, all the way down the length of the bed. Plant seeds by hand, three inches apart in each furrow. About one ounce or 500 seeds will be needed to plant three furrows at this spacing in a bed that is five feet wide and fifty feet long.

Cover the seeds with 1 1/2 inches of topsoil. After planting, carefully step down each row to firm the soil around the seeds. To finish planting, rake one inch of leaves back over the bed as mulch. After a couple of rainstorms no one will be able to detect that any planting has occurred. The site should look completely natural.

The stratified seed will germinate next spring. The plants will look like three small strawberry leaves on a stem about an inch tall. Some of the seeds will not germinate, and some will be eaten by rodents. Over the next seven years, the plant population in each bed will be reduced every year by various natural forces. The final stand will most likely be a thin healthy population of wild ginseng plants.

In the wild-simulated method, after planting, little or no work is required until harvest 7-10 years later. The ginseng plants are left to the mercy of nature. Weeds will compete with the plants for water and nutrients, the plants will struggle to grow in barely cultivated soil. These stressful conditions result in the wild appearance of the roots that are eventually harvested. Digging the roots is difficult because the roots are entangled with other woodland plants. Harvested roots should be air dried in the shade.

A half acre will require 10 pounds of seeds and will yield anywhere from 0 to 200 pounds of dried root in 6 to 10 years.

The greatest threat is theft and crops should not be grown in areas where people dig wild ginseng. It is best to plant on controlled land and to keep quiet about the enterprise. Growing ginseng away from the mountains where most ginseng hunters search is also a good way to keep the crop safe. Someone may come across it but probably would not know what it was.

Ideal growing conditions for ginseng are more difficult to find in low-lying regions than they are in the mountains. The forest floor in most woodland areas is too hot and dry during the summer for ginseng to thrive. (Moist environments may be found that are good, if not perfect, places to grow ginseng.) Small pockets of cooler soil may be found very often on a north facing hillside above a stream or river. Many landowners with the right conditions are successfully growing ginseng well out of the mountains.

Henna

Henna (*Lawsonia inermis*, also known as hina, the henna tree, the mignonette tree, and the Egyptian privet) is a flowering plant and the sole species of the *Lawsonia* genus.

The name *henna* also refers to the dye prepared from the plant and the art of temporary tattooing based on those dyes. Henna has been used since antiquity to dye skin, hair, and fingernails, as well as fabrics including silk, wool, and leather. The name is used in other skin and hair dyes, such as *black henna* and *neutral henna*, neither of which is derived from the henna plant.

Historically, henna was used for cosmetic purposes in Ancient Egypt, as well as other parts of North Africa, West Africa, the Horn of Africa, the Arabian Peninsula, the Near East and South Asia. It was also popular among women in Iberia and 19th-century Europe. Today, bridal henna nights remain an important tradition in many of these areas.

Description

Henna is a tall shrub or small tree, standing 1.8 to 7.6 m tall (6 to 25 ft). It is glabrous and multi-branched, with spine-tipped branchlets. The leaves grow opposite each other on the stem and are glabrous, sub-sessile, elliptical, and lanceolate (long and wider in the middle; average dimensions are 1.5–5.0 cm x 0.5–2 cm or 0.6–2 in x 0.2–0.8 in), acuminate (tapering to a long point), and have depressed veins on the dorsal surface. Henna flowers have four sepals and a 2 mm (0.079 in) calyx tube, with 3 mm (0.12 in) spread lobes. Its petals are obvate, with white or red stamens found in pairs on the rim of the calyx tube. The ovary is four-celled, 5 mm (0.20 in) long, and erect. Henna fruits are small, brownish capsules, 4–8 mm (0.16–0.31 in) in diametre, with 32–49 seeds per fruit, and open irregularly into four splits.

Cultivation

The henna plant is native to northern Africa, western and southern Asia, and northern Australasia, in semi-arid zones and tropical areas. It produces the most dye when grown in temperatures between 35 and 45 °C (95 and 113 °F). During the onset of precipitation intervals, the plant grows rapidly, putting out new shoots. Growth subsequently slows. The leaves gradually yellow and fall during prolonged dry or cool intervals. It does not thrive where minimum temperatures are below 11 °C (52 °F). Temperatures below 5 °C (41 °F) will kill the henna plant.

India has seven climate regions, based on rainfall and temperature. Northern areas near the Himalayas are very cold, and southern areas are tropical. The east has heavy rainfalls and the west is dry. Each area has different agricultural practices reflecting these differences. Henna is an ideal crop for some regions, and impossible in others.

India's agricultural cycle is created by the monsoon system, with planting and harvesting driven by summer monsoon rainfall. The eastern arm of India has heavy monsoon rainfall, as high as 10,900mm annually, almost 10 times that of New York City, and the highest in the world. The central southern interior averages about 650mm per year, but varies regionally, with about the same rain amounts as are across France. Rajasthan gets less rain than any other part of India, and the western edge is a sand desert. Rajasthan's henna growing regions get less than 400mm precipitation per year, comparable to the mesa areas of New Mexico. When the monsoons fail Rajasthan, they bring 250mm rain or less to the arid farming regions. In these drought years, henna crops may be smaller, but have higher dye content. If henna grows in soils that are moist it is attacked by *Corticium koleroga* (Black rot disease) and by *Xanthomonas lawsonieae* (Bacterial leafspot), which damage the crop and diminish yield. If henna is grown in soils that are dry, or have long periods of drought, this is rarely a problem.

With global warming, summer highs have increased all across India. Farmers in hot, arid regions find henna a reliable cash crop when drought and heat withers other crops, and river levels drop too low for irrigation. Hotter, dryer weather, with prolonged droughts favour high quality henna. Henna is a small tree, and stays in the ground for many years through harvest after harvest, with no need for plowing. It holds the fragile dry topsoil in place, helping maintain productive farmland. Henna also tolerates soils salinized by previous irrigation.

Henna grows through most of India, though it is a major commercial crop only in the northwestern states where the climate is relatively hot and dry. Gujarat, Madhya Pradesh and Punjab farm henna, primarily to be sold as hair dye. Rajasthani henna farms often produce body art quality henna.

Gujarat averages 500 mm to 700 mm rainfall per year in the northern part of the state, 2000 mm per year in the southern part. In the summer temperatures average between 25° and 43° C (77° and 109° F), but during heat waves, can rise to 48° C (118° F). The dryer, hotter areas of Gujarat produce good quality henna.

Madhya Pradesh has more rainfall, with 1000mm in the west to 2000mm annually in the east. Summer temperatures range from 33C to 44C, which henna tolerates well, though some areas have winter minimums too cool for henna to thrive. Though Madhya Pradesh produces henna, most of it is harvested for hair dye, or the flowers for perfume oils, rather than body art quality henna.

Many areas of the Punjab have January frosts, which will kill a henna plant, so only sheltered or southern areas have henna. Summer highs are frequently over 35C, and as high as 45C. Rainfall ranges between less than 500mm annually, near the desert area, and 2000mm in the hilly area. Some regions of Punjab are suitable for henna cultivation, but other areas are inhospitable.

The Henna Industry in Rajasthan

Rajasthan's western region gets less than 400 mm rainfall annually, and the semi-arid Pali region gets 400m -500mm, unless the monsoon fails. 90% all rain in Pali falls during the monsoon, in two months between late June and late August, so most of the district is in permanent agricultural drought. The area cannot dependably support rain fed agriculture, and if rivers are very low, there isn't enough electricity generated at the dams to irrigate. Farmers who are assured irrigation can grow wheat, cotton, mustard, tobacco, chilies, and millet, crops common to Rajasthan. However, irrigation will always gradually salinize soil, and plowing can destroy fragile, dry soils. In severe droughts, no farmer can be assured irrigation. Henna farming has the advantage that it needs neither irrigation nor plowing.

The pre-monsoon season, the months prior to June, is the hottest season, with temperatures of 35 to 48 C or more. This region is vulnerable to the Thar Desert encroaching from west of Jodhpur, especially during the pre-monsoon season, when dust storms from the west carry away loose topsoil. Henna, a perennial with a tenacious root system, preserves farmland. The best henna harvests are at the end of this dry season, as the clouds gather and the monsoon begins, when dye levels peak from heat and drought and new dye-intense growth bursts forth. Henna may be harvested 2, 3 or 4 times a year, with the early July crop being the best, and the February crop being the weakest.

Sojat and the surrounding areas in the Pali district of Rajasthan, are the most intensive henna cultivation areas in India. Up to 90 percent of Rajasthan's commercial henna production is in this area. Farmers in this region have produced henna for generations, and regard it as a desirable crop, as well as part of their heritage. Much of this area is hilly, with small creeks and some rivers. Henna cuttings can be rooted on riverbanks, and then transplanted to fields after their first flowering. When they are in the fields, they grow as stunted trees, pruned 2 to 4 times a year. Henna leaves are dried, the flowers are gathered to be pressed for perfume oil, and the twigs fed to cattle. Henna requires little or no fertilizer, survives on rainwater, and requires little specialized labour. Both men and women work the fields. Women are often responsible for weeding and harvesting; men take care of soil preparation and transportation. A family can often manage their own cultivation and harvest. Henna will survive a drought season that would ruin other crops, and is therefore highly attractive to farmers.

Farmers have increased their henna plantings each year since, with the Asian Development Bank lending money and expertise to farmers so they can increase production, to raise production over a tonne per hectare, and to reduce labour. The world demand for high quality henna is increasing, both for safe hair dyes, and for body art quality henna.

The Indian henna export cities of Udaipur, Delhi, Jodphur, Jaipur, Calcutta, Mumbai (Bombay) and Ahamadabad have had an advantage over other henna producing countries such as Iran and Yemen. Indian business often uses English as a second language, so they are able to correspond with clients in the west with little misunderstanding. These cities have relatively stable internet hosting, and can call upon competent web designers, to display their products. Shipments from India do not undergo the intensive security checks as do those from Iran and Yemen. India has relatively good diplomatic relations with the west, and this facilitates economic and communication flow. India has a banking platform that enables them to conduct financial transactions with the west. These infrastructure advantages have allowed the west to have relatively easy henna transactions with India, as opposed to poor or near impossible transactions with Iran, Yemen, Nigeria, Mauritania, and Sudan, where farmers also produce high quality henna. This advantage is so great, that many western people believe henna is produced ONLY in India, simply because that is the most accessible henna.

Preparation and Application

Body art:

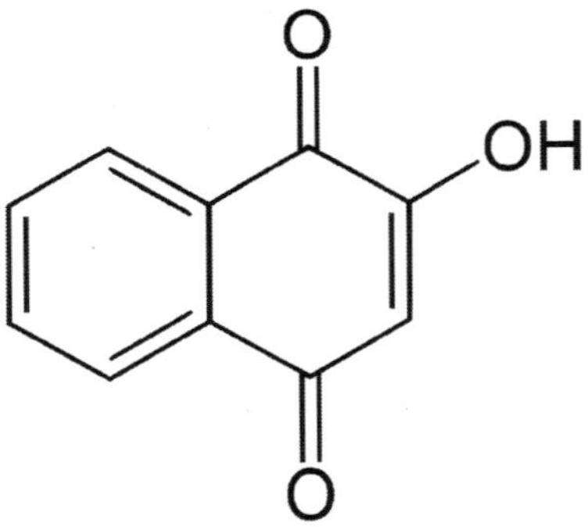

***Figure:** Lawsone, an active compound in Henna*

Whole, unbroken henna leaves will not stain the skin. Henna will not stain skin until the lawsone molecules are made available (released) from the henna leaf. Fresh henna leaves will stain the skin if they are smashed with a mildly acidic liquid. The lawsone will gradually migrate from the henna paste into the outer layer of the skin and bind to the proteins in it, creating a fast stain.

Since it is difficult to form intricate patterns from coarse crushed leaves, henna is commonly traded as a powder made by drying, milling and sifting the leaves. The dry powder is mixed with lemon juice, strong tea, or other mildly acidic liquids to make a preparation with toothpaste-like consistency, which can be used to make finely detailed body art. The henna mix must rest for 6 to 24 hours before use, to release the lawsone from the leaf matter. Essential oils with high levels of monoterpene alcohols, such as tea tree, eucalyptus, cajeput, or lavender, will improve skin stain characteristics.

The paste can be applied with many traditional and innovative tools, including resist, a cone, syringe, Jac bottle or fingers. A light stain may be achieved within minutes, the longer the paste is left on the skin, the stronger the stain will be, and should be left for several hours. To prevent it from drying or falling off the skin, the paste is often sealed down by dabbing a sugar/lemon mix over the dried paste, or simply adding some form of sugar to the paste. It is debatable whether this adds to the colour of the end result; some believe it increases the intensity of the shade. After time the dry paste is simply brushed or scraped away.

Henna stains are orange soon after application, but darken over the following three days to a reddish brown. Soles and palms have the thickest layer of skin and so take up the most lawsone, and take it to the greatest depth, so that hands and feet will have the darkest and most long-lasting stains. Steaming or warming the henna pattern will darken the stain, either during the time the paste is still on the skin, or after the paste has been removed. Chlorinated water and soaps may spoil the darkening process: alkaline products may hasten the darkening process. After the stain reaches its peak colour it will appear to fade, as the stained dead cells exfoliate.

Hair Dye

History: Henna has been used as a cosmetic hair dye for 6,000 years. In Ancient Egypt, it is known to have been worn. Henna has also traditionally been used for centuries in other parts of North Africa, the Horn of Africa, the Arabian Peninsula, the Near East and South Asia.

In Ancient Egypt, Ahmose-Henuttamehu (17th Dynasty, 1574 BCE): Henuttamehu was probably a daughter of Seqenenre Tao and Ahmose Inhapy. Smith reports that the mummy of Henuttamehu's own hair had been dyed a bright red at the sides, probably with henna.

In Europe, henna was popular among women connected to the aesthetic movement and the Pre-Raphaelite artists of England in the 1800s. Dante Gabriel Rossetti's wife and muse, Elizabeth Siddal, had naturally bright red hair. Contrary to the cultural tradition in Britain that considered red hair unattractive, the Pre-Raphaelites fetishized red hair. Siddal was portrayed by Rossetti in many paintings that emphasized her flowing red hair. The other Pre-Raphaelites, including Evelyn De Morgan and Frederick Sandys, academic classicists such as Frederic Leighton, and French painters such as Gaston Bussière and the Impressionists further popularized the association of henna-dyed hair and young bohemian women.

Opera singer Adelina Patti is sometimes credited with popularizing the use of henna in Europe in the late 1800s. Parisian courtesan Cora

Pearl was often referred to as La Lune Rousse (the red moon) for dying her hair red. In her memoirs, she relates an incident when she dyed her pet dog's fur to match her own hair. By the 1950s, Lucille Ball popularized "henna rinse" as her character, Lucy Ricardo, called it on the television show I Love Lucy. It gained popularity among young people in the 1960s through growing interest in Eastern cultures.

Muslim men may use henna as a dye for hair and most particularly their beards. This is considered *sunnah*, a commendable tradition of the Prophet Muhammad. Furthermore, a *hadith* (narration of the Prophet) holds that he encouraged Muslim women to dye their nails with henna to demonstrate femininity and distinguish their hands from the hands of men; thus some Muslim women in the Middle East apply henna to their finger and toenails as well as their hands.

Today

Commercially packaged henna, intended for use as a cosmetic hair dye, is available in many countries, and is now popular in India, as well as the Middle East, Europe, Australia, Canada and the United States. The colour that results from dying with henna depends on the original colour of the hair, as well as the quality of the henna, and can range from orange to auburn to burgundy. Henna can be mixed with other natural hair dyes including Cassia Obovata for lighter shades of red or even blond, or with indigo to achieve brown and black shades. Some products sold as "henna" include these other natural dyes. Others may include metal salts that can interact with other chemical treatments, or oils and waxes that may inhibit the dye, or even chemical dyes which are common allergens. Any product that comes in a cream, block, or paste form has some sort of additives.

As with henna in body art, the dried leaf powder should be mixed with a mild acid such as lemon juice, orange juice, or vinegar and left to stand. The resulting paste is then applied to the hair, and covered with plastic wrap to keep it from drying out. This paste should be left in the hair for several hours in order for the dye to permanently bind to the hair strands. The paste is then washed away leaving hair that is permanently dyed. Sometimes henna is mixed with hot or boiling water and used immediately. This gives a colour that may fade, and which is not as rich or deep.

Traditions of Henna as Body Art

The different words for henna in ancient languages imply that it had more than one point of discovery and origin, as well as different pathways of daily and ceremonial use.

Henna has been used to adorn young women's bodies as part of social and holiday celebrations since the late Bronze Age in the eastern Mediterranean. The earliest text mentioning henna in the context of marriage and fertility celebrations comes from the Ugaritic legend of Baal and Anath, which has references to women marking themselves with henna in preparation to meet their husbands, and Anath adorning herself with henna to celebrate a victory over the enemies of Baal.

Wall paintings excavated at Akrotiri (dating prior to the eruption of Thera in 1680 BCE) show women with markings consistent with henna on their nails, palms and soles, in a tableau consistent with the henna bridal description from Ugarit. Many statuettes of young women dating between 1500 and 500 BCE along the Mediterranean coastline have raised hands with markings consistent with henna. This early connection between young, fertile women and henna seems to be the origin of the Night of the Henna, which is now celebrated worldwide.

The Night of the Henna was celebrated by most groups in the areas where henna grew naturally: Jews, Muslims, Sikhs, Hindus, Christians and Zoroastrians, among others, all celebrated marriages by adorning the bride, and often the groom, with henna. Across the henna-growing region, Purim, Eid, Diwali, Karva Chauth, Passover, Nowruz, Mawlid, and most saints' days were celebrated with some henna. Favourite horses, donkeys, and salukis had their hooves, paws, and tails hennaed

Battle victories, births, circumcision, birthdays, Zâr, as well as weddings, usually included some henna as part of the celebration. When there was joy, there was henna, as long as henna was available.

Henna was regarded as having Barakah ("blessings"), and was applied for luck as well as joy and beauty. Brides typically had the most henna, and the most complex patterns, to support their greatest joy, and wishes for luck. Some bridal traditions were very complex, such as those in Yemen, where the Jewish bridal henna process took four or five days to complete, with multiple applications and resist work. The fashion of "Bridal Mehndi" in Pakistan, Northern Libya and in North Indian diasporas is currently growing in complexity and elaboration, with new innovations in glitter, gilding, and fine-line work. Recent technological innovations in grinding, sifting, temperature control, and packaging henna, as well as government encouragement for henna cultivation, have improved dye content and artistic potential for henna.

Though traditional henna artists were Nai caste in India, and barbering castes in other countries (lower social classes), talented contemporary henna artists can command high fees for their work. Women in countries where women are discouraged from working outside the home can find socially acceptable, lucrative work doing henna. Morocco, Mauritania, Yemen, Libya, Somalia, Sudan, as well as India and many other countries have thriving women's henna businesses. These businesses are often open all night for Eid, Diwali and Karva Chauth. Many women may work together during a large wedding, wherein hundreds of guests have henna applied to their body parts. This particular event at a marriage is known as the Mehndi Celebration or Mehndi Night, and is mainly held for the bride and groom.

Regions

Bridal henna nights are a popular tradition in North Africa, the Horn of Africa, the Arabian Peninsula, the Near East and South Asia.

Algeria

In Algeria, the bride's mother-in-law presents her with jewelry and paints the henna on her hands.

India

In India, the longer the henna stays on the bride's hand the longer it is believed that her in-laws will treat her well. If the henna fades out quickly, it's a sign that she will not be happily married.

Saudi Arabia

In Saudi Arabia, one of the female relatives of the bride is the one who paints the bride's hands with henna, but the relative has to be happily married or else she will bring bad luck to the bride.

Israel

In Israel, Mizrahi and Sephardic Jewish women sometimes choose to have a hennah party about a week before the wedding. The hennah party is smaller than the wedding, as only closer friends and family members are invited. The bride and groom wear traditional costumes as do some of the main guests. There is much dancing and music, especially when the hennah is brought out. The hennah is usually presented in a deep dish with lit candles in it and carried by the grandmother. She applies the hennah onto the palms of the bride and groom and they are blessed. Subsequently, guests stain their palms with hennah as well. There are variations in customs and dress between the different Jewish communities (Yemenite, Moroccan, Indian, etc.).

Somalia

In Somalia, henna is worn by Somali women on their hands, arms, feet and neck during weddings, Eid, Ramadan, and other festive occasions. Somali henna designs are similar to those in the Arabian peninsula, often featuring flower motifs and triangular shapes. The palm is also frequently decorated with a dot of henna and the fingertips are dipped in the dye. Henna parties are usually held before the wedding takes place.

Tunisia

In Tunisia, henna celebrations last for seven days. On the 3rd day, the bride wears a traditional dress and has henna painted on her hands and feet. As for the groom, his pinky finger is painted with henna on the 6th day.

Turkey

In Turkey, henna is sold in convenience stores and markets. Among these are the Egyptian Bazaar in Istanbul.

Saffron

Saffron is a spice derived from the flower of *Crocus sativus*, commonly known as the saffron crocus. *Crocus* is a genus in the family Iridaceae. Saffron crocus grows to 20–30 cm (8–12 in) and bears up to four flowers, each with three vivid crimson stigmas, which are the distal end of a carpel. Together with the styles, or stalks that connect the stigmas to their host plant, the dried stigmas are used mainly in various cuisines as a seasoning and colouring agent. Saffron, long among the world's most costly spices by weight, is native to Greece or Southwest Asia and was first cultivated in Greece. As a genetically monomorphic clone, it was slowly propagated throughout much of Eurasia and was later brought to parts of North Africa, North America, and Oceania.

Figure: *Saffron – valuable stigmas, or threads, are tediously plucked, piled, and dried.*

The saffron crocus, unknown in the wild, probably descends from *Crocus cartwrightianus*, which originated in Crete; *C. thomasii* and *C.*

pallasii are other possible precursors. The saffron crocus is a triploid that is "self-incompatible" and male sterile; it undergoes aberrant meiosis and is hence incapable of independent sexual reproduction—all propagation is by vegetative multiplication via manual "divide-and-set" of a starter clone or by interspecific hybridisation. If *C. sativus* is a mutant form of *C. cartwrightianus*, then it may have emerged via plant breeding, which would have selected for elongated stigmas, in late Bronze Age Crete.

Saffron's taste and iodoform- or hay-like fragrance result from the chemicals picrocrocin and safranal. It also contains a carotenoid dye, crocin, which imparts a rich golden-yellow hue to dishes and textiles. Its recorded history is attested in a 7th-century BC Assyrian botanical treatise compiled under Ashurbanipal, and it has been traded and used for over four millennia. Iran now accounts for approximately 90% of the world production of saffron.

Description

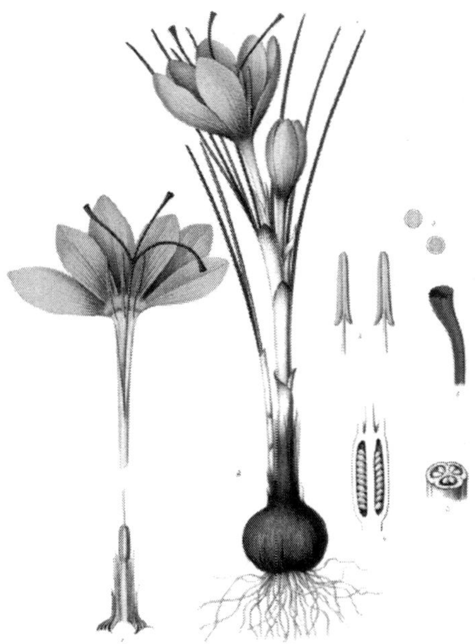

Figure: *Köhler's Medicinal Plants*

The domesticated saffron crocus, *Crocus sativus*, is an autumn-flowering perennial plant unknown in the wild. Its progenitors are possibly the eastern Mediterranean autumn-flowering *Crocus cartwrightianus*, which is also known as "wild saffron" and originated in Greece. The saffron crocus probably resulted when *C. cartwrightianus* was subjected to extensive artificial selection by growers seeking longer stigmas. *C. thomasii* and *C. pallasii* are other possible sources.

It is a sterile triploid form, which means that three homologous sets of chromosomes compose each specimen's genetic complement; *C. sativus* bears eight chromosomal bodies per set, making for 24 in total. Being sterile, the purple flowers of *C. sativus* fail to produce viable seeds; reproduction hinges on human assistance: clusters of corms, underground, bulb-like, starch-storing organs, must be dug up, divided, and replanted. A corm survives for one season, producing via this vegetative division up to ten "cormlets" that can grow into new plants in the next season. The compact corms are small, brown globules that can measure as large as 5 cm (2.0 in) in diametre, have a flat base, and are shrouded in a dense mat of parallel fibres; this coat is referred to as the "corm tunic". Corms also bear vertical fibres, thin and net-like, that grow up to 5 cm above the plant's neck.

The plant grows to a height of 20–30 cm (8–12 in), and sprouts 5–11 white and non-photosynthetic leaves known as cataphylls. These membrane-like structures cover and protect the crocus's 5 to 11 true leaves as they bud and develop. The latter are thin, straight, and blade-like green foliage leaves, which are 1–3 mm in diametre, either expand after the flowers have opened ("hysteranthous") or do so simultaneously with their blooming ("synanthous"). *C. sativus* cataphylls are suspected by some to manifest prior to blooming when the plant is irrigated relatively early in the growing season. Its floral axes, or flower-bearing structures, bear bracteoles, or specialised leaves that sprout from the flower stems; the latter are known as pedicels. After aestivating in spring, the plant sends up its true leaves, each up to 40 cm (16 in) in length. In autumn, purple buds appear. Only in October, after most other flowering plants have released their seeds, do its brilliantly hued flowers develop; they range from a light pastel shade of lilac to a darker and more striated mauve. The flowers possess a sweet, honey-like fragrance. Upon flowering, plants average less than 30 cm (12 in) in height. A three-pronged style emerges from each flower. Each prong terminates with a vivid crimson stigma 25–30 mm (0.98–1.18 in) in length.

Cultivation

Crocus sativus thrives in the Mediterranean maquis, an ecotype superficially resembling the North American chaparral, and similar climates where hot and dry summer breezes sweep semi-arid lands. It can nonetheless survive cold winters, tolerating frosts as low as "10 °C (14 °F) and short periods of snow cover. Irrigation is required if grown outside of moist environments such as Kashmir, where annual rainfall averages 1,000–1,500 mm (39–59 in); saffron-growing regions in Greece

(500 mm or 20 in annually) and Spain (400 mm or 16 in) are far drier than the main cultivating Iranian regions. What makes this possible is the timing of the local wet seasons; generous spring rains and drier summers are optimal. Rain immediately preceding flowering boosts saffron yields; rainy or cold weather during flowering promotes disease and reduces yields. Persistently damp and hot conditions harm the crops, and rabbits, rats, and birds cause damage by digging up corms. Nematodes, leaf rusts, and corm rot pose other threats. Yet *Bacillus subtilis* inoculation may provide some benefit to growers by speeding corm growth and increasing stigma biomass yield.

The plants fare poorly in shady conditions; they grow best in full sunlight. Fields that slope towards the sunlight are optimal (i.e., south-sloping in the Northern Hemisphere). Planting is mostly done in June in the Northern Hemisphere, where corms are lodged 7–15 cm (2.8–5.9 in) deep; its roots, stems, and leaves can develop between October and February. Planting depth and corm spacing, in concert with climate, are critical factors in determining yields. Mother corms planted deeper yield higher-quality saffron, though form fewer flower buds and daughter corms. Italian growers optimise thread yield by planting 15 cm (5.9 in) deep and in rows 2–3 cm (0.79–1.18 in) apart; depths of 8–10 cm (3.1–3.9 in) optimise flower and corm production. Greek, Moroccan, and Spanish growers employ distinct depths and spacings that suit their locales.

C. sativus prefers friable, loose, low-density, well-watered, and well-drained clay-calcareous soils with high organic content. Traditional raised beds promote good drainage. Soil organic content was historically boosted via application of some 20–30 tonnes of manure per hectare. Afterwards, and with no further manure application, corms were planted. After a period of dormancy through the summer, the corms send up their narrow leaves and begin to bud in early autumn. Only in mid-autumn do they flower. Harvests are by necessity a speedy affair: after blossoming at dawn, flowers quickly wilt as the day passes. All plants bloom within a window of one or two weeks. Roughly 150 flowers together yield but 1 g (0.035 oz) of dry saffron threads; to produce 12 g (0.42 oz) of dried saffron (or 72 g (2.5 oz) moist and freshly harvested), 1 kg (2.2 lb) of flowers are needed; 1 lb (0.45 kg) yields 0.2 oz (5.7 g) of dried saffron. One freshly picked flower yields an average 30 mg (0.0011 oz) of fresh saffron or 7 mg (0.00025 oz) dried.

Spice

Chemistry: Saffron contains more than 150 volatile and aroma-yielding compounds. It also has many nonvolatile active components, many of which are carotenoids, including zeaxanthin, lycopene, and various α- and β-carotenes. However, saffron's golden yellow-orange colour is primarily the result of α-crocin. This crocin is trans-crocetin di-(β-D-gentiobiosyl) ester; it bears the systematic (IUPAC) name 8,8-diapo-8,8-carotenoic acid. This means that the crocin underlying saffron's aroma is a digentiobiose ester of the carotenoid crocetin. Crocins themselves are a series of hydrophilic carotenoids that are either monoglycosyl or diglycosyl polyene esters of crocetin. Crocetin is a conjugated polyene dicarboxylic acid that is hydrophobic, and thus oil-soluble. When crocetin is esterified with two water-soluble gentiobioses, which are sugars, a product results that is itself water-soluble. The resultant α-crocin is a carotenoid pigment that may comprise more than 10% of dry saffron's mass. The two esterified gentiobioses make α-crocin ideal for colouring water-based and non-fatty foods such as rice dishes.

The bitter glucoside picrocrocin is responsible for saffron's flavour. Picrocrocin (chemical formula: C16H26O7; systematic name: 4-(β-D-glucopyranosyloxy)-2,6,6- trimethylcyclohex-1-ene-1-carboxaldehyde) is a union of an aldehyde sub-element known as safranal (systematic name: 2,6,6-trimethylcyclohexa-1,3-diene-1-carboxaldehyde) and a carbohydrate. It has insecticidal and pesticidal properties, and may comprise up to 4% of dry saffron. Picrocrocin is a truncated version of the carotenoid zeaxanthin that is produced via oxidative cleavage, and is the glycoside of the terpene aldehyde safranal. The reddish-coloured zeaxanthin is, incidentally, one of the carotenoids naturally present within the retina of the human eye.

When saffron is dried after its harvest, the heat, combined with enzymatic action, splits picrocrocin to yield D–glucose and a free safranal molecule. Safranal, a volatile oil, gives saffron much of its distinctive aroma. Safranal is less bitter than picrocrocin and may comprise up to 70% of dry saffron's volatile fraction in some samples. A second element underlying saffron's aroma is 2-hydroxy-4,4,6-trimethyl-2,5-cyclohexadien-1-one, which produces a scent described as saffron, dried hay-like. Chemists find this is the most powerful contributor to saffron's fragrance, despite its presence in a lesser quantity than safranal. Dry saffron is highly sensitive to fluctuating pH levels, and rapidly breaks down chemically in the presence of light and oxidising agents. It must, therefore, be stored away in air-tight containers to minimise contact with atmospheric oxygen. Saffron is somewhat more resistant to heat.

Grades and ISO 3632 Categories

Saffron is not all of the same quality and strength. Strength is related to several factors including the amount of style picked along with the red stigma. Age of the saffron is also a factor. More style included means the saffron is less strong gram for gram, because the colour and flavour are concentrated in the red stigmas. Saffron from Iran, Spain and Kashmir is classified into various grades according to the relative amounts of red stigma and yellow styles it contains. Grades of Iranian saffron are: "sargol" (red stigma tips only, strongest grade), "pushal" or "pushali" (red stigmas plus some yellow style, lower strength), "bunch" saffron (red stigmas plus large amount of yellow style, presented in a tiny bundle like a miniature wheatsheaf) and "konge" (yellow style only, claimed to have aroma but with very little, if any, colouring potential). Grades of Spanish saffron are "coupé" (the strongest grade, like Iranian sargol), "mancha" (like Iranian pushal), and in order of further decreasing strength "rio", "standard" and "sierra" saffron. The word "mancha" in the Spanish classification can have two meanings: a general grade of saffron or a very high quality Spanish-grown saffron from a specific geographical origin. Real Spanish-grown La Mancha saffron has PDO protected status and this is displayed on the product packaging. Spanish growers fought hard for Protected Status because they felt that imports of Iranian saffron re-packaged in Spain and sold as "Spanish Mancha saffron" were undermining the genuine La Mancha brand.

Countries producing less saffron do not have specialised words for different grades and may only produce one grade. Artisan producers in Europe and New Zealand have offset their higher labour charges for saffron harvesting by targeting quality, only offering extremely high grade saffron.

In addition to descriptions based on how the saffron is picked, saffron may be categorised under the international standard ISO 3632 after laboratory measurement of crocin (responsible for saffron's colour), picrocrocin (taste), and safranal (fragrance or aroma) content. However, often there is no clear grading information on the product packaging and little of the saffron readily available in UK is labelled with ISO category. This lack of information makes it hard for customers to make informed choices when comparing prices and buying saffron.

Under ISO 3632, determination of non-stigma content ("floral waste content") and other extraneous matter such as inorganic material ("ash") are also key. Grading standards are set by the International Organization for Standardization, a federation of national standards bodies. ISO 3632 deals exclusively with saffron and establishes three categories: III (poorest quality), II, and I (finest quality). Formerly there was also category IV, which was below category III. Samples are assigned categories by gauging the spice's crocin and picrocrocin content, revealed by measurements of specific spectrophotometric absorbance. Safranal is treated slightly differently and rather than there being threshold levels for each category, samples must give a reading of 20-50 for all categories.

These data are measured through spectrophotometry reports at certified testing laboratories worldwide. Higher absorbances imply greater levels of crocin, picrocrocin and safranal, and thus a greater colouring potential and therefore strength per gram. The absorbance reading of crocin is known as the "colouring strength" of that saffron. Saffron's colouring strength can range from lower than 80 (for all category IV saffron) up to 200 or greater (for category I). The world's finest samples (the selected, most red-maroon, tips of stigmas picked from the finest flowers) receive colouring strengths in excess of 250, making such saffron over three times more powerful than category IV saffron. Market prices for saffron types follow directly from these ISO categories. Sargol and coupé saffron would typically fall into ISO 3632 category I. Pushal and mancha would probably be assigned to category II. On many saffron packaging labels, neither the ISO 3632 category nor the colouring strength (the measurement of crocin content) is displayed.

However, many growers, traders, and consumers reject such lab test numbers. Some people prefer a more holistic method of sampling batches of threads for taste, aroma, pliability, and other traits in a fashion similar to that practised by experienced wine tasters. However, ISO 3632 grade and colouring strength information allow consumers to make instant comparisons between the quality of different saffron brands, without needing to purchase and sample the saffron. In particular, consumers can work out value for money based on price per unit of colouring strength rather than price per gram, given the wide possible range of colouring strengths that different kinds of saffron can have.

Despite attempts at quality control and standardisation, an extensive history of saffron adulteration, particularly among the cheapest grades, continues into modern times. Adulteration was first documented in Europe's Middle Ages, when those found selling adulterated saffron were executed under the *Safranschou* code. Typical methods include mixing in extraneous substances like beets, pomegranate fibres, red-dyed silk fibres, or the saffron crocus's tasteless and odourless yellow stamens. Other methods included dousing saffron fibres with viscid substances like honey or vegetable oil to increase their weight. However, powdered saffron is more prone to adulteration, with turmeric, paprika, and other powders used as diluting fillers. Adulteration can also consist of selling mislabelled mixes of different saffron grades. Thus, in India, high-grade Kashmiri saffron is often sold and mixed with cheaper

Iranian imports; these mixes are then marketed as pure Kashmiri saffron, a development that has cost Kashmiri growers much of their income.

Types

Figure: *Saffron from different producer countries, picked and dried in different ways gives rise to different end qualities.*

The various saffron crocus cultivars give rise to thread types that are often regionally distributed and characteristically distinct. Varieties (not varieties in the botanical sense) from Spain, including the tradenames "Spanish Superior" and "Creme", are generally mellower in colour, flavour, and aroma; they are graded by government-imposed standards. Italian varieties are slightly more potent than Spanish. The most intense varieties tend to be Iranian. Various "boutique" crops are available from New Zealand, France, Switzerland, England, the United States, and other countries—some of them organically grown. In the U.S., Pennsylvania Dutch saffron—known for its "earthy" notes—is marketed in small quantities.

Consumers may regard certain cultivars as "premium" quality. The "Aquila" saffron, or *zafferano dell'Aquila*, is defined by high safranal and crocin content, distinctive thread shape, unusually pungent aroma, and intense colour; it is grown exclusively on eight hectares in the Navelli Valley of Italy's Abruzzo region, near L'Aquila. It was first introduced to Italy by a Dominican monk from Inquisition-era Spain. But the biggest saffron cultivation in Italy is in San Gavino Monreale, Sardinia, where it is grown on 40 hectares, representing 60% of Italian production; it too has unusually high crocin, picrocrocin, and safranal content. Another is the "Mongra" or "Lacha" saffron of Kashmir (*Crocus sativus* 'Cashmirianus'), which is among the most difficult for consumers to obtain. Repeated droughts, blights, and crop failures in the Indian-controlled areas of Kashmir combine with an Indian export ban to contribute to its prohibitive overseas prices. Kashmiri saffron is recognisable by its dark maroon-purple hue; it is among the world's darkest, which hints at strong flavour, aroma, and colouring effect.

Saffron Cultivation in Kashmir, India

Undoubtedly, India has made great strides in the field of social and economic development since independence. No one can deny the fact that agriculture has an important role to play in the economy of Jammu and Kashmir. Eighty per cent population of this state lives in villages and depends on agriculture.

The state can not progress unless the economic condition of this vast majority improves and the standard of living of people gets enhanced. Agriculture itself is a vast field. It's like a tree which has endless branches. One among them is related to saffron cultivation.

Saffron has several names-Zafran, Kesar, Kang, Kang Posh etc. Kang posh, the flowers of Saffron is a symbol of freshness and purity. The vast stretches of saffron fields give the impression of a newly wedded bride draped in a saffron shawl taking a nap. Kashmir is known as the valley of flowers. Among several varieties of flowers grown here, saffron has its own importance and utility.

Historically, the cultivation of saffron started around three or four centuries back in Arabia and Spain. Thereafter, its cultivation spread as far as Iran, Sweden and India. Now, saffron growing is a great commercial activity. In Indian agriculture, this activity is also known as "Golden Zest".

The cultivation of saffron is a traditional art. In India, 5,707 hectares of land comes under its cultivation. Its annual production is around sixteen thousand kilograms. The state of Jammu and Kashmir tops the list of the saffron growing states in India. it can be gauged from the fact that out of the total 5,707 hectares of land under its cultivation 4,496 hectares lie exclusively in Jammu and Kashmir.

In Kashmir, Pampore, which is situated at a distance of 15 kilometres from Srinagar, is famous the world over for its high grade saffron. Saffron is also grown, though in a limited scale, in Kishtwar of Jammu. Pampore and its neighbouring areas produce an average of 2,128 kilograms of saffron every year.

Sir Water Lawrence, in his book "The Valley of Kashmir" has written about saffron. He undertook a detailed study of the cultivation of saffron and its utilities. He writes people from different parts of Kashmir used to thromg Pampore for the cultivation of Saffron. Now, this is the exclusive preserve of the local population. Increasingly, saffron then used to be sold at a price of one rupee a tola. He adds that till 1923, the maximum production of saffron was recorded in the Naga area of Pampore. Now the average earning from its cultivation is from 30 to 40 crore rupees. Hence, after the production of fruits, the cultivation of saffron is the second largest activity in the state. So, it is necessary that new techniques are used to promote the cultivation of saffron. Modern and scientific techniques must be used.

This is incumbent upon the state government and the agriculture scientists that they develop self confidence among the saffron growers. They must take sufficient interest in their education and training and encourage them. They must also help them in using the new techniques to preserve their produce.

The difficulties saffron growers have to face normally in the state is not only that the production is less then the investment, but the diseases are also a big problem. Scientifically, this is critical to make proper arrangement for proper diagnosis of several diseases found in saffron trees. Several diseases are common in saffron such as Corm Rot, Dry Rot, Root Rot, Bacteria Rot, Ring Rot, Charcoal Rot, Mosaic etc.

Corm rot among these diseases is perceived as most deadly. A few years back, a survey was conducted in Pulwama district of Kashmir in this regard. As a result, it was found that the disease was more prevalent in certain villages. To contain this and other diseases the agriculture experts must be contacted and their advice must be put to good use. The expert of the Sher-e-Kashmir University of Agriculture Science and Technology could rightly guide the saffron growers. By seeking their expert views, the saffron plants could be saved from the onslaught of different diseases. Its production could also be enhanced.

In Jammu and Kashmir, saffron is normally sown in August and till the 15 of September. The flowers are plucked in October and November. During harvesting or plucking of flowers, the atmosphere must not be hot. So, this process is completed early in the morning. From sunrise to 10 AM is perceived the ideal time for flower plucking. After harvesting, the flowers are kept for drying for 5 days. Afterwards, they are kept in an airy container so that the quality of the produce does not deteriorate. Now-a-days solar drier is also used. It takes only seven, eight hours to dry the produce.

Locally, one kilogram saffron consists of around one lakh sixty thousand to one lakh seventy thousand tiny flowers. It's so tiring a job, and time consuming too.

To enhance the production of saffron a few things must be considered:

- Provision of good quality and high yield seeds.
- Extension in the area of cultivation and production.
- Extensive use of solar drier and air drier. Financial help for the purchase of these apparatus.
- Appropriate training for packing the produce.
- The expert's total and equal cooperation and coordination with the saffron growers.

Marketing is also one of the main problems. An average saffron grower finds it troublesome to sell the small quantity of his produce. The grading and packing individually is not highly profitable. The cultivator has scarce resources. So, this is necessary that cooperative societies are formed to sell the saffron at remunerative prices. A proper strategy must be devised and the whole activity must be so organized that the average saffron growers gets full benefits of his toils.

Some Other Important Medicinal Plants

Marijuana

Seriously. Though marijuana is still illegal in the United States, it is legal in 12 states for medicinal purposes, and if a case of poison ivy in the woods isn't a medicinal purpose, what is? Marijuana was *mostly* legal until 1970 when it became classified as a hard drug. No one thought of it as a dangerous or illicit drug until the 20th century; in fact, hemp was George Washington's primary crop and Thomas Jefferson's secondary crop. The Declaration of Independence is written on it; the Gutenberg Bible was printed on hemp, too. There's actually an environmental dimension to legalizing marijuana – hemp is a remarkable and renewable plant, offering all kinds of foodstuff and product uses that surpass cotton and plastic. But health benefits are well documented, from depression and anxiety relief to reduced blood pressure, pain alleviation and glaucoma treatment. It is not addictive, does not kill brain cells and is not a "gateway" drug – in fact, when pot is more available, studies show that the use of hard drugs like heroin and cocaine actually decreases. The bottom line for hikers: when your leg is broken from a misjudged boulder hopping attempt (pain) and a bear has eaten your friend (depression) and you're lost because you forgot the compass (dumbass), consult the cannabis.

Cannabis, commonly known as marijuana and by numerous other names, is a preparation of the cannabis plant intended for use as a psychoactive drug and as medicine. Pharmacologically, the principal psychoactive constituent of cannabis is tetrahydrocannabinol (THC); it is one of 483 known compounds in the plant, including at least 84 other cannabinoids, such as cannabidiol (CBD), cannabinol (CBN), tetrahydrocannabivarin (THCV), and cannabigerol (CBG). Cannabis is often consumed for its psychoactive and physiological effects, which can include heightened mood or euphoria, relaxation, and an increase in appetite. Unwanted side-effects can sometimes include a decrease in short-term memory, dry mouth, impaired motor skills, reddening of the eyes, and feelings of paranoia or anxiety.

Contemporary uses of cannabis are as a recreational or medicinal drug, and as part of religious or spiritual rites; the earliest recorded uses date from the 3rd millennium BC. Since the early 20th century cannabis has been subject to legal restrictions with the possession, use, and sale of cannabis preparations containing psychoactive cannabinoids currently illegal in most countries of the world; the United Nations has said that cannabis is the most-used illicit drug in the world. In 2004, the United Nations estimated that global consumption of cannabis indicated that approximately 4% of the adult world population (162 million people) used cannabis annually, and that approximately 0.6% (22.5 million) of people used cannabis daily.

Marijuana consists of the dried flowers and subtending leaves and stems of the female *Cannabis* plant. This is the most widely consumed form, containing 3% to 22% THC. In contrast, cannabis varieties used to produce industrial hemp contain less than 1% THC and are thus not valued for recreational use. This is the stock material from which all other preparations are derived. It is noted that cannabis or its extracts must be sufficiently heated or dehydrated to cause decarboxylation of its most abundant cannabinoid, tetrahydrocannabinolic acid (THCA), into psychoactive THC.

Kief

Kief is a powder, rich in trichomes, which can be sifted from the leaves and flowers of cannabis plants and either consumed in powder form or compressed to produce cakes of hashish. The word "kif" derives from Arabic: ßíÝ *kayf*, meaning *well-being* or *pleasure*.

Hashish

Hashish (also spelled hasheesh, hashisha, or simply hash) is a concentrated resin cake or ball produced from pressed kief, the detached trichomes and fine material that falls off of cannabis flowers and leaves. It varies in colour from black to golden brown depending upon purity and variety of cultivar it was obtained from. It can be consumed orally or smoked.

Tincture

Cannabinoids can be extracted from cannabis plant matter using high-proof spirits (often grain alcohol) to create a tincture, often referred to as "green dragon". Nabiximols is a branded product name from a tincture manufacturing pharmaceutical company.

Hash Oil

Hash oil is obtained from the cannabis plant by solvent extraction, and contains the cannabinoids present in the natural oils of cannabis flowers and leaves. The solvents are evaporated to leave behind a very concentrated oil. Hemp oil is very different from both hemp seed oil and cannabis flower essential oil. Owing to its purity, these products are consumed by smoking, vaporizing, eating, or topical application.

Infusions

There are many varieties of cannabis infusions owing to the variety of non-volatile solvents used. The plant material is mixed with the solvent and then pressed and filtered to express the oils of the plant into the solvent. Examples of solvents used in this process are cocoa butter, dairy butter, cooking oil, glycerine, and skin moisturizers. Depending on the solvent, these may be used in cannabis foods or applied topically.

Adulterated Cannabis

Contaminants may be found in hashish obtained from "soap bar"-type sources. The dried flowers of the plant may be contaminated by the plant taking up heavy metals and other toxins from its growing environment, or by the addition of glass. In the Netherlands, chalk has been used to make cannabis appear to be of a higher quality. Increasing the weight of hashish products in Germany with lead caused lead intoxication in at least 29 users. Despite cannabis being generally perceived as a natural or chemical-free product, in a recent Australian survey one in four Australians consider cannabis grown indoors under hydroponic conditions to be a greater health risk due to increased contamination, added to the plant during cultivation to enhance the plant growth and quality.

Production: It is often claimed by growers and breeders of herbal cannabis that advances in breeding and cultivation techniques have increased the potency of cannabis since the late 1960s and early '70s, when THC was first discovered and understood. However, potent seedless cannabis such as "Thai sticks" were already available at that time. *Sinsemilla* (Spanish for "without seed") is the dried, seedless inflorescences of female cannabis plants. Because THC production drops off once pollination occurs, the male plants (which produce little THC themselves) are eliminated before they shed pollen to prevent pollination. Advanced cultivation techniques such as hydroponics, cloning, high-intensity artificial lighting, and the sea of green method are frequently employed as a response (in part) to prohibition enforcement efforts that make outdoor cultivation more risky. It is often cited that the average levels of THC in cannabis sold in United States rose dramatically between the 1970s and 2000, but such statements are likely skewed because of undue weight given to much more expensive and potent, but less prevalent samples.

"Skunk" refers to several named strains of potent cannabis, grown through selective breeding and sometimes hydroponics. It is a cross-breed of *Cannabis sativa* and *C. indica* (although other strains of this mix exist in abundance). Skunk cannabis potency ranges usually from 6% to 15% and rarely as high as 20%. The average THC level in coffee shops in the Netherlands is about 18–19%.

Cannabis Cultivation

Cannabis belongs to the genus *Cannabis* in the family Cannabaceae and includes three species, *C. sativa*, *C. indica*, and *C. ruderalis* (APG II system). It is typically a dioecious (each individual is either male or female) annual plant (life period: April–September).

C. sativa and *C. indica* generally grow tall (some varieties reach 4 m) and their females start the production of flowers rich in THC (1% to 29%) as the season changes from summer to autumn. *C. ruderalis* is very short, produces only trace amounts of THC, but is very rich in CBD and flowers independently of the photoperiod and according to age. However, commercial cross-bred hybrids containing both ruderalis, indica and/or sativa genes exist (usually called autoflowering).

Cultivation Requirements: Cannabis needs certain conditions to flourish.

Growth Medium: Soil is required, except for cannabis grown with hydroponics or aeroponics

- Sufficient nutrients—commercial potting soils usually indicate this as "N-P-K = x%-y%-z%" the percentages of the fundamental nutritional elements, i.e., nitrogen, phosphorus and potassium. Nutrients are often provided to the soil via fertilizers but such practice requires caution.
- pH between 5.9 and 6.5. This value can be adjusted.

Warmth: The optimal day temperature range for cannabis is 24 to 30 °C (75 to 86 °F). Temperatures above 31 °C (88F) and below 15.5 °C (60F) seem to decrease THC potency and slow growth. At 13 °C the plant undergoes a mild shock, though some strains withstand frost temporarily. Frost occurs when air temperatures dip below 0 °C (32 °F) and ice crystals form.

Light: Light can be natural (outdoor growing) or artificial (indoor growing).

Under artificial light, the plant typically remains under a regime of 16–20 hours of light and 4–8 hours of darkness from the germination until flowering, although the plant can use a full 24 hours of light without harm. There is an ongoing debate over the importance of the "dark period". It has been shown that, when subjected to a regimen of constant light without a dark period, most types of flora, including Cannabis, will begin to show signs of decreased photosynthetic response, lack of vigor, and an overall decrease in vascular development. There is no defined flowering "stage", unless you are discussing an auto-flowering variety (Cannabis Ruderalis). Typically, flowering is induced by changing the light schedule to 12 hours of light and 12 hours of darkness. Flowering in cannabis is triggered by a hormonal reaction within the plant that is initiated by an increase in its dark cycle, i.e. the plant needs sufficient prolonged darkness for bract/bracteole (flowering) to begin. Some Indica varieties require as little as 8 hours of dark to begin flowering, whereas some Sativa varieties require up to 13 hours.

Water: Watering frequency and amount is determined by many factors, including temperature and light, the age, size and stage of growth of the plant and the medium's ability to retain water. A conspicuous sign of water problems is the downward wilting of leaves. Too much water can kill young cannabis plants.

The additives in municipal water can reduce plant growth and can be removed with reverse osmosis filtration but is expensive and produces poorer results than well or spring water.

Nutrients: Nutrients are taken up from the soil by plants. Nutrient soil amendments (fertilizers) are added when the soil nutrients are depleted. Fertilizers can be chemical or organic, liquid or powder, and usually contain a mixture of ingredients. Commercial fertilizers indicate the levels of NPK (nitrogen, phosphorus, and potassium). During the vegetative stage, cannabis needs more N than P and K, while during the flowering stage, P is more essential than N and K. The presence of secondary nutrients (calcium, magnesium, sulfur) is recommended.

Micro nutrients (e.g. iron, boron, chlorine, manganese, copper, zinc, molybdenum) rarely manifest as deficiencies.

Because Cannabis' nutrient needs vary widely depending on the variety, they are usually determined by trial and error and fertilizers are applied sparingly to avoid burning the plant.

Stages of Development: *Germination:* Germination is the process in which the seeds sprout and the root emerges. In Cannabis it takes from 12 hours to 8 days. Warmth, darkness and moisture initiate metabolic processes such as the activation of hormones that trigger the expansion of the embryo within the seed. Then the seed coat cracks open and a small embryonic root emerges and begins growing downward (because of gravitropism), if placed in a proper growing medium. Soon (after 2–4 days) the root is anchored and two circular embryonic leaves (cotyledons) emerge in search of light and the remains of the seed shell are pushed away. This marks the beginning of the seedling stage.

Germination is initiated by soaking seeds either between wet paper towels, in a cup of water at room temperature, in wet peat pellets, or directly in potting soil. Peat pellets are often used as a germinating medium because the saturated pellets with their seedlings are planted directly into the intended growing medium with a minimum of shock to the plant.

Seedling Phase: The seedling stage begins when the seed coat splits open and exposes the root and round "seed leaves" or cotyledons. It lasts from 1 to 4 weeks and is the period of greatest vulnerability in the life cycle of the plant, requiring moderate humidity levels, medium to high light intensity, and adequate but not excessive soil moisture. Most indoor growers use compact fluorescent or T5 fluorescent lights during this stage as they produce little heat. HPS and MH lights produce large amounts of radiant heat and increase the rate of transpiration in the plant which can quickly dry out seedlings with their small root systems.

The plant will naturally begin to develop identifiable sex characteristics in this stage after 4 to 6 weeks, but some growers will hasten this by switching to a 12/12 hour light period to induce flowering and, once sex is determined and the male plants are removed, returning the plants to the vegetation stage with an 18/6 hour light period. However, forcing a plant to flower and then return to the vegetative stage adds significant extra growing time and has been shown to produce lower yields and reduce potencies by a factor of 2 to 3, so growers who wish to sex the plants as early as possible without these drawbacks take a clone of each seedling, root it, and force flowering in a separate growing area.

Vegetative Phase: Duration: 1–2 months indoors. In this stage the plant needs all the light (at least 18 hours) and nutrients (food) that it can use, depending on the genetics of the particular plant. It continues to grow vertically and produce new leaves. The sex is starting to reveal itself, which is a sign that the next stage begins. Concurrently the root system expands downwards in search of more water and food. Some newly developed strains (auto flowering hybrids) omit the vegetative stage and pass directly from seedling to pre-flowering.

When the plant possesses seven sets of true leaves and the 8th is barely visible in the centre of the growth tip, or shoot apical meristem (SAM), the plant has entered the vegetative phase of growth. During the vegetative phase, the plant directs its energy resources primarily to the growth of leaves, stems, and roots. A strong root system is required for strong floral development. A plant needs 1 or 2 months to mature before blooming. The plant is ready when it has revealed its sex. Plant size is a good indicator of sex. Females tend to be shorter and branchier due to their raceme type inflorescence than males, whose flowers grow in panicles. The males tend to have fewer leaves near the top and females have more leaves surrounding the flowers. The males are then usually culled when they are identified, so that the females will not be pollinated, thus producing "sin semilla" ("without seed") buds.

During the vegetative phase, cultivators generally employ an 18- to 24-hour photoperiod because the plants grow more quickly if they receive more light, although a warmer and cooler period are required for optimal health. Although no dark period is required, there is debate among cultivators as to whether a dark period is beneficial, and many continue to employ a dark period. Energy savings often support using a dark period, as plants undergo late day decline and therefore lighting during the late night hours is less effective.

The amount of time to grow a cannabis plant indoors in the vegetative stage depends on the size of the flower, the light used, the size of the space, and how many plants are intended to flower at once, and how big the strain gets in "the stretch" (i.e., the first two weeks of flowering).

Cannabis cultivators employ fertilizers high in N (nitrogen) and K (potassium) during the vegetative stage, as well as a complete micro nutrient fertilizer. The strength of the fertilizer is gradually increased as the plants grow and become more hardy.

The modification of a plant's growth habit is called training. Indoor cultivators employ many training techniques to encourage shorter plants and denser canopy growth. For example, unless the crop is too large to be extensively pruned, cultivators remove adventitious growth shoots, often called suckers, that are near the bottom of the plant and/or receive little light and will produce poor quality buds. Many cultivators also employ other techniques:

Topping: Topping is the removal of the top of the apical meristem (dominant central stem), called the apex or terminal bud, to transfer apical dominance (the tendency for the apex to grow more rapidly than the rest of the plant) to the shoots emanating from the two nodes immediately beneath the pruning cut. This process can be repeated on one or both of the two new meristems, when they become apically dominant, with the same results. This process can actually be repeated nigh infinitely, but over-diffusion of apical dominance produces smaller, lower quality buds, so it is usually done no more than a few times. Topping also causes more rapid growth of all of the branches below the cut while the plant heals.

Pinching: Pinching (also called super cropping) is similar to topping in that it causes lower branches to grow more rapidly, but the apical meristem maintain apical dominance, which is especially useful if the plant has already been topped. Pinching is performed by firmly pinching the apical meristem(s) so as to substantially damage vascular and structural cells but without totally breaking the stem. This causes lower limbs to grow more rapidly while the pinched tissue heals, after which time the stem resumes apical dominance.

LST'ing

LST stands for *L*ow *S*tress *T*raining and is another form of supercropping, many times referred to as LST super-cropping. This technique involves bending and tying the plants branches to manipulate the plant into a more preferred growth shape. This method of training works very well for indoor growers who need to illuminate their plants using overhead lights. Since light intensity greatly diminishes with increased distance (Inverse-square law), LST'ing can be used to keep all growth tips (meristems) at the same distance from the light and can achieve optimal light exposure. LST can be used in conjunction with topping, since topping increases axial growth (side shoots). Topping is often done a few weeks before beginning LST'ing. The training works by changing the distribution of hormones—more specifically Auxins—in the plant. LST'ing resembles the training of grape vines into their support lattices. Outdoor gardeners also employ training techniques to keep their plants from becoming too vertical.

Pre-flowering Phase

Also called the *stretch*, this takes one day to two weeks. Most plants last for 10–14 days after switching the light cycle to 12/12. Plant development increases dramatically, with the plant doubling or more in size. Production of more branches and nodes occurs in this stage, as the structure for flowering grows. The plant starts to show bracts/ bracteoles where the branches meet the stem (nodes). Pre-flowering indicates the plant is ready to flower.

Flowering Phase: Varies from about 6 to 22 weeks with pure indicas having a shorter flowering time than pure sativas. Mixed indica/ sativa strains have an intermediate flowering time. The sex is clearly revealed in the first two weeks of 12/12. Males produce little balls clustered together like grapes. Most plants (except auto flowering strains that flower independently of photoperiod) can flower under diminished light. In nature, cannabis plants sense the forthcoming winter as the Earth revolves around the Sun and daylight reduces in duration. If females are not pollinated (fertilized by male pollen) they start to produce buds that contain sticky white resin glands or trichomes in a final attempt for pollination by windborne male pollen. The trichomes produce resins containing the largest amounts of THC and CBD, the two main psychoactive substances. Fertilized females produce resinous trichomes but more plant energy is sunk into seeds, which can be half the mass of a fertilized inflorescence; thus, to maximize resin per gram, seedless cultivation is preferred.

Buds of the first case are called *sin semilla* (which translates to "without seeds" in Spanish, and is often misspelled as one word). Potent sin semilla is especially important to medical users, to minimize the amount of cannabis they must consume to be afforded relief. Cannabis with seeds is generally considered to be of inferior quality and/or grown with inferior technique.

Indoors, cannabis is induced into flowering by decreasing its photoperiod to at least 10 hours of darkness per day. In order to initiate a flowering response, the number of hours of darkness must exceed a critical point. Generally the more hours of darkness each day, the shorter the overall flowering period but the lower the yield. Conversely, the fewer hours of darkness each day, the longer the overall flowering period and the higher the yield. Traditionally most growers change their plants lighting cycle to 12 hours on and 12 hours off since this works as a happy medium that most strains respond well to. This change in photoperiod mimics the plant's natural outdoor cycle, with up to 18 hours of light per day in the summer and down to less than 12 hours of light in fall and winter. Some 'semi-autoflowering' strains that have been bred exclusively for outdoor use, particularly in outdoor climates such as that of the UK, the plants will start flowering with as much as 16–17 hours of light per day. Usually they can start flowering in July and finish far earlier than other strains, particularly those that haven't been bred as outdoor strains. Because the weather in northern latitudes gets very wet and cold during October when many strains start to finish flowering, these semi-autoflowering strains give an advantage as they are far less likely to suffer from botrytis (grey mold) which is caused by wet weather. Alternatively growers may artificially induce the flowering period during the warmer months by blacking out the plants for 12 hours a day i.e. by covering the plants with black plastic for example, which excludes all light during this period so the plant can flower even during long days.

Although the flowering hormone in most plants (including cannabis) is present during all phases of growth, it is inhibited by exposure to light. To induce flowering, the plant must be subject to at least 8 hours of darkness per day; this number is very strain-specific and most growers flower with 12 hours of darkness to be safe.

Flowers from certain plants (e.g. cannabis) are called bract/bracteole, and are (with cannabis) the most prized part of the plant. In late flowering the bract/bracteole are easily visible to the naked eye. Bract/bracteole development begins approximately 1–2 weeks after the photoperiod is reduced. In the first weeks of flowering a plant usually doubles in size and can triple. Bract/bracteole development ends around 5 weeks into flowering and is proceeded by a period of bract/bracteole "swelling". During this time the buds greatly increase in weight and size.

Outdoor Cannabis Cultivation: Cannabis can be planted outdoors, either on natural soil or in pots of pre-made or commercial soil. Some strains perform better than others in outdoor settings, an attribute that depends on different conditions, variables and aspects. Outdoor marijuana strains, like most other strains, can be bought in numerous locations and over a hundred different cannabis strains that are bred for outdoor growing exist—many of these outdoor cannabis seeds are simply copies of other pre-existent strains or seeds with different names and descriptions.

To generate optimum quantities of THC-containing resin the plant needs a fertile soil and long hours of daylight. This means THC production for outdoor growth occurs optimally anywhere within 35° of the equator. Typical growing regions include Mexico, Nepal, Northern India, many parts of Africa, Afghanistan, The United States and Australia. In most places of the subtropics, cannabis is germinated from late spring to early summer and harvested from late summer to early autumn.

Outdoor cultivation is common in both rural and urban areas. Outdoor cultivators tend to grow indica-based strains because of its heavy yields, quick maturing time, and shortness. Some growers prefer sativa because of its clear-headed (cerebral) high, better response to sunlight, and lower odor emissions.

One can cultivate on his own property or practice guerrilla farming i.e. plant cannabis in remote areas such as forest clearings or mountain cliffs and visit rarely. However, such a method is prone to ripoffs – so much so that some ingenious growers even attach pots to trees to decrease this possibility. Guerrilla growing has given birth to the activist movement Operation Overgrow, where the plant is grown with the explicit purpose of introducing the cannabis plant into the natural ecosystem. For outdoor cultivation, growers choose areas that receive twelve hours or more of sunlight a day. In the Northern Hemisphere, growers typically plant seeds in late May or early June to provide plants a full four months of growth. Harvest is usually between mid-September and early October. In North America, northern locations are preferred (North Coast of California and British Columbia being particularly notable), but southern locations (such as Maui, Hawaii) are also known to be good producers. Where local laws do not permit growing cannabis, cultivators sometimes grow in forests or rugged and rural areas where the local population is unlikely to find the crop. Cannabis is also grown hidden by a crop that is taller, such as maize. This is reported by the United States government to be common in the midwestern states. Bamboo and elderberry are also used as camouflage companion plants.

Indoor Cannabis Cultivation: Cannabis can be grown indoors in a soil-like medium under artificial light, adding fertilizer when the plants are given water. Cultivating cannabis indoors is more complicated and expensive than growing outdoors, but it allows the cultivator complete control over the growing environment. Plants of any type can be grown faster indoors than out due to 24-hour light, additional atmospheric CO_2, and controlled humidity which allows freer CO_2 respiration. Plants can also be grown indoors through the use of hydroponics.

To grow plants indoors, a growing medium (e.g. soil or growing substrate), water, nutrients, light and air need to be supplied to the plant (with the exception of aeroponic cultivation, in which case a growing medium is not required).

Advanced Cultivation Methods: These methods include:

1. using a water or air-based growth medium (known as hydroponics and aeroponics respectively)
2. the use of homemade, organic composted fertilizers
3. training and trellising techniques such as *Screen of Green* (also known as *SCROG*), *Sea of Green* (also known as *SOG*) "Super

cropping" and LST super cropping; and entire systems and methods such as the *NIMBY no-dump* method, *Hempy Bucket*, and the *Krusty Freedom Bucket* methods. Research into the production of cannabis for the drug Marinol and other more profitable and marketable forms of cannabis based medicines has further pushed the envelope of cannabis cultivation in all forms of laboratory, both public and private.

The emphasis on advanced cultivation techniques, as well as the availability of hybrid strains (with names like *Northern Lights*, *Master Kush*, *NYC Diesel*), is believed to be a factor in the increase in the overall quality and variety of commercially-available cannabis over the past few decades. The internet in particular has brought together widely diverse genetics from around the world through trading and purchasing. However, well-grown heirloom strains (e.g. island sweet skunk, fruity Thai etc.) are used to produce 1 gram per watt harvest.

Lady Ferns

If you grew up in the Pacific Northwest you likely know what ferns are good for: treating stinging nettles. One of the world's oldest plants, there are many varieties of ferns, but if you're lucky enough to spy the soft, delicate lady fern, grab some and roll it up between your palms into a rough mash. The juices released will quickly ease stinging nettle burns and can also ease minor cuts, stings and burns (fresh salt water also works in a pinch for bee stings). Bracken fern are similar to lady fern and will work, as well. The rougher, glossier, stiff sword fern and deer fern won't be as effective, though. Lady ferns actually grow all over North America but are common in areas with high rainfall.

Athyrium filix-femina (lady fern or common lady-fern) is a large, feathery species of fern native throughout most of the temperate Northern Hemisphere, where it is often abundant (one of the more common ferns) in damp, shady woodland environments and is often grown for decoration.

Its common names "lady fern" and "female fern" refer to how its reproductive structures (sori) are concealed in an inconspicuous — deemed "female" — manner on the frond. Alternatively, it is said to be feminine because of its elegant and graceful appearance.

Athyrium filix-femina is now commonly split into two species, *A. angustum* (narrow lady fern) and *A. asplenioides* (southern lady fern). The southern lady fern has a broader frond, especially at the base.

Both species are cespitose (the fronds arising from a central point as a clump rather than along a rhizome). The deciduous fronds are light yellow-green, 20-90 cm long and 5-25 cm broad. Sori appear as dots on the underside of the frond, 1-6 per pinnule. They are covered by a prominently whitish to brown reniform (kidney-shaped) indusium. Fronds are very dissected, being 3-pinnate. The stipe may bear long, pale brown, papery scales at the base. The spores are yellow on *A. angustum* and dark brown on *A. asplenioides*.

Cultivation and Uses: This plant has gained the Royal Horticultural Society's Award of Garden Merit. The rhizomes and young fronds are poisonous when raw, but edible after cooking.

How to Grow Lady Fern: Lady ferns, also known as Athyrium filix-femina, are hardy ferns common to U.S. Department of Agriculture Hardiness Zones 3 to 8. Several varieties of lady ferns exist that can grow up to 5 wide and high, but more commonly they grow about 2 to 3 feet in height and width. They can generally grow without much care.

Step 1: Select an area that is in partial shade and with soil that is both well draining and slightly acidic. Add humus to your soil before planting to help with drainage and improve acidity.

Step 2: Plant in the spring, after the last frost. Dig a hole that is just as deep as the container, but twice as wide. If transplanting rhizomes (the plant's bulb-like root), dig a hole that is just deep enough for the rhizome but wide enough to fan out its roots.

Step 3: Space multiple plants at least 2 feet apart. Since there are several varieties of lady ferns, follow label instructions. You will most likely thin out the plants in later years, so the closer together you plant them, the sooner you will need to thin them out.

Step 4: Keep the soil moist throughout the spring and summer months. Some lady ferns can tolerate dry soil, so read the label for watering instructions specifically for your plant.

California Poppy

The brilliant blooms of the poppy make this opioid plant an iconic one. The plant is an effective nervine (anxiety reliever) and is safe for use on agitated children. Can be made into a a tea for quick relief of nervousness and tension. A stronger decoction will offer pain relief. (A decoction is made by "stewing" all safe plant parts, including stems and roots if possible, in water for several hours and, ideally, soaking overnight.)

Eschscholzia californica (California poppy, golden poppy, California sunlight, cup of gold) is a species of flowering plant in the family Papaveraceae, native to the United States and Mexico, and the official state flower of California.

It is a perennial or annual growing to 5–60 in (13–152 cm) tall, with alternately branching glaucous blue-green foliage. The leaves are ternately divided into round, lobed segments. The flowers are solitary on long stems, silky-textured, with four petals, each petal 2 to 6 cm (0.79 to 2.36 in) long and broad; flower colour ranges from yellow to orange, with flowering from February to September. The petals close at night or in cold, windy weather and open again the following morning, although they may remain closed in cloudy weather. The fruit is a slender, dehiscent capsule 3 to 9 cm (1.2 to 3.5 in) long, which splits in two to release the numerous small black or dark brown seeds. It survives mild winters in its native range, dying completely in colder climates.

Habitat: Its native habitat includes California, extending to Oregon, Washington, Nevada, Arizona, New Mexico, Sonora and northwest Baja California. The Antelope Valley California Poppy Reserve is located in northern Los Angeles County, California. At the peak of the blooming season, orange petals seem to cover all 1,745 acres (706 ha) of the reserve. Other prominent locations of California poppy meadows are in Bear Valley (California, Colusa County), Point Buchon and numerous other locations.

Taxonomy: The species is highly variable, with over 90 synonyms. Some botanists accept two subspecies — one with four varieties (e.g., Leger and Rice, 2003) — though others do not recognize them as distinct (e.g., Jepson 1993):

- *E. californica* subsp. *californica*, native to California, Baja California, and Oregon, widely planted as an ornamental, and an invasive elsewhere.
    - *E. californica* subsp. *californica* var. *californica*, which is found along the coast from the San Francisco Peninsula north. They are perennial and somewhat prostrate, with yellow flowers.
    - *E. californica* subsp. *californica* var. *maritima* (E. L. Greene) Jeps., which is found along the coast from Monterey south to San Miguel Island. They are perennial, long-lived, glaucous, short in stature, and have extremely prostrate growth and yellow flowers.
    - *E. californica* subsp. *californica* var. *crocea* (Benth.) Jeps., which grows in non-arid inland regions. They are perennial, taller, and have orange flowers.
    - *E. California* subsp. *californica* var. *peninsularis* (E. L. Greene) Munz, which is an annual or facultative annual growing in arid inland environments.

- *E. californica* subsp. *mexicana* (E. L. Greene) C. Clark, the Mexican Gold Poppy, which is found in the Sonoran Desert. Some authorities refer to it as *E. Mexicana.*

Cultivation: *E. californica* is drought-tolerant, self-seeding, and easy to grow in gardens. It is best grown as an annual, in full sun and sandy, well-drained, poor soil. Horticulturalists have produced numerous cultivars with a range of colours and blossom and stem forms. These typically do not breed true on reseeding. Seeds are often sold as mixtures. The following cultivars have gained the Royal Horticultural Society's Award of Garden Merit:-

- *E. californica*
- 'Appleblossom Bush' (pink)
- 'Dali' (red)
- 'Lemon Bush' (pale yellow)
- 'Rose Chiffon' (pink and white)
- Thai Silk Series (mixed colours)

Other Uses: California poppy leaves were used medicinally by Native Americans, and the pollen was used cosmetically. The seeds are used in cooking.

An aqueous extract of the plant has sedative and anxiolytic action. The extract acts as a mild sedative when smoked. The effect is far milder than that of opium. California poppy contains a different class of alkaloids:

> *"An aqueous alcohol extract of* Eschscholzia californica *has been evaluated for benzodiazepine, neuroleptic, antidepressant, antihistaminic and analgesic properties. The plant extract did not protect mice against the convulsant effects of pentylenetetrazol, and did not cause muscle relaxant effects, but appeared to possess an affinity for the benzodiazepine receptor. The extract induced peripheral analgesic effects in mice but did not possess antidepressant, neuroleptic or antihistaminic effects."*

Invasive Species: Because of its beauty and ease of growing, the California poppy was introduced into several regions with similar Mediterranean climates. It is commercially sold and widely naturalized in Australia, and was introduced to South Africa, Chile, and Argentina. It is recognized as a potentially invasive species within the United States, although no indications of ill effects have been reported for this plant where it has been introduced outside of California. Ironically, it has been displaced in large areas of its original habitat, such as Southern California, by more invasive exotic species, such as mustard or annual grasses.

Chilean Population: In Chile, it was introduced from multiple sources between the mid-19th century and the early 20th century. It appears to have been both intentionally imported as an ornamental garden plant, and accidentally introduced along with alfalfa seed grown in California. Since Chile and California have similar climatic regions and have experienced much agricultural exchange, it is perhaps not surprising that it was introduced to Chile. Once there, its perennial forms spread primarily in human-disturbed environments.

Interestingly, the introduced Chilean populations of California poppy appear to be larger and more fecund in their introduced range than in their native range. Introduced populations have been noted to be larger and more reproductively successful than native ones (Elton, 1958), and there has been much speculation as to why. Increase in resource availability, decreased competition, and release from enemy pressure have all been proposed as explanations. One hypothesis is that the resources devoted in the native range to a defense strategy, can in the absence of enemies be devoted to increased growth and reproduction. However, this is not the case with introduced populations of *E. californica* in Chile: the Chilean populations were actually more resistant to Californian caterpillars than the native populations.

State Flower: As the official state flower of California *Eschscholzia californica* is pictured on welcome signs along highways entering California. It was selected as the state flower by the California State Floral Society in December 1890, winning out over the Mariposa lily (genus *Calochortus*) and the Matilija poppy (*Romneya coulteri*) by a landslide, but the state legislature did not make the selection official until 1903. Its golden blooms were deemed a fitting symbol for the Golden State. April 6 is designated California Poppy Day.

Blood Flower: The blood flower (also Mexican butterfly weed) is a type of tropical milkweed with toxic milky sap that is emetic (it makes you hurl). It's also historically favoured as a heart stimulant and worm expellent. Pretty useful for a number of potential hiking disasters, if you think about it. (Of course, if you'd quit eating those poisonous berries you probably wouldn't need to worry about finding a natural expectorant.)

Typical plants are evergreen perennial subshrubs that grow up to 1 m (3.3 ft) tall and have pale gray stems. The leaves are arranged oppositely on the stems and are lanceolate or oblong-lanceolate shaped ending in acuminate or acute tips. Like other members of the genus, the sap is milky. The flowers are in cymes with 10-20 flowers each. They have purple or red corollas and corona lobes that are yellow or orange. Flowering occurs nearly year round. The 5–10 cm (2.0–3.9 in) long, fusiform shaped fruits are called follicles. The follicles contain tan to brown seeds that are ovate in shape and 6–7 mm (0.24–0.28 in) long. The flat seeds have silky hairs that allow the seeds to float on air currents when the pod-like follicles dehisce (split open).

Cultivation: There are a number of different cultivars with improved flower colours and shorter habit, some have brilliant red, yellow or orange coloured flowers. *Asclepias curassavica* is excellent in butterfly gardens or as a cut flower. However, when the stems or leaves are broken, a poisonous milky sap exudes which can cause eye injury.

Distribution: *Asclepias curassavica* is described by NatureServe as a "widespread species, ranging from southern North America through Central America and into South America." It is an introduced species in the US states of California, Florida, Hawaii, Louisiana, Tennessee, and Texas, as well as the US unincorporated territories of Puerto Rico and the United States Virgin Islands.

It has been introduced and naturalized in the Chinese provinces of Anhui, Fujian, Guangdong, Guangxi, Guizhou, Hainan, Hubei, Hunan, Jiangsu, Jiangxi, Qinghai, Sichuan, Xizang, Yunnan, and Zhejiang, as well as in Taiwan. It is considered an exotic plant, but not a weed, at the Meteor Downs South Project near Rolleston, Queensland, Australia.

Tansy

If you've decided to backpack through Europe instead of the mountains of Mexico (but why?), you'll want to know about a few helpful medicinal plants. Tansy is an old-world aster and remedy, used for flavoring beer and stews as well as repelling insects. Rubbing the leaves on the skin provides an effective bug repellent, but tansy can also be used to treat worms. It is said to be poisonous when extracted, but a few leaves are not harmful if ingested.

Tansy (*Tanacetum vulgare*) is a perennial, herbaceous flowering plant of the aster family, native to temperate Europe and Asia. It has been introduced to other parts of the world and in some areas has become invasive. It is also known as Common Tansy, Bitter Buttons, Cow Bitter, Mugwort, or Golden Buttons.

Tansy is a flowering herbaceous plant with finely divided compound leaves and yellow, button-like flowers. It has a stout, somewhat reddish, erect stem, usually smooth, 50–150 cm tall, and branching near the top.

The leaves are alternate, 10–15 cm long and are pinnately lobed, divided almost to the centre into about seven pairs of segments, or lobes, which are again divided into smaller lobes having saw-toothed edges, giving the leaf a somewhat fernlike appearance. The roundish, flat-topped, button-like, yellow flower heads are produced in terminal clusters from mid-to-late summer. The scent is similar to that of camphor with hints of rosemary. The leaves and flowers are toxic if consumed in large quantities; the volatile oil contains toxic compounds including thujone, which can cause convulsions and liver and brain damage. Some insects, notably the tansy beetle *Chrysolina graminis*, have resistance to the toxins and subsist almost exclusively on the plant.

History and Distribution: Tansy is native to Eurasia; it is found in almost all parts of mainland Europe. It is absent from Siberia and some of the Mediterranean islands. The ancient Greeks may have been the first to cultivate it as a medicinal herb. In the sixteenth century it was considered to be "necessary for a garden" in Britain.

History of Uses: Tansy has a long history of use. It was first recorded as being cultivated by the ancient Greeks for medicinal purposes. In the 8th century AD it was grown in the herb gardens of Charlemagne and by Benedictine monks of the Swiss monastery of Saint Gall. Tansy was used to treat intestinal worms, rheumatism, digestive problems, fevers, sores, and to "bring out" measles.

During the Middle Ages and later, high doses were used to induce abortions. Contradictorily, tansy was also used to help women conceive and to prevent miscarriages. In the 15th century, Christians began serving tansy with Lenten meals to commemorate the bitter herbs eaten by the Israelites. Tansy was thought to have the added Lenten benefits of controlling flatulence brought on by days of eating fish and pulses and of preventing the intestinal worms believed to be caused by eating fish during Lent.

Tansy was used as a face wash and was reported to lighten and purify the skin. In the 19th century, Irish folklore suggested that bathing in a solution of tansy and salts would cure joint pain. Although most of its medicinal uses have been discredited, tansy is still a component of some medicines and is listed by the United States Pharmacopeia as a treatment for fevers, feverish colds, and jaundice.

Insect Repellent: Tansy has also been cultivated and used for its insect repellent and in the worm warding type of embalming. It was packed into coffins, wrapped in funeral winding sheets, and tansy wreaths were sometimes placed on the dead. Henry Dunster, the first president of Harvard University, was buried wearing a tansy wreath in a coffin packed with tansy; when "God's Acre" was moved in 1846 the tansy had maintained its shape and fragrance, helping to identify the president's remains. By the 19th century, tansy was used so much at New England funerals that people began to disdain it for its morbid association with death.

During the American colonial period, meat was frequently rubbed with or packed in tansy leaves to repel insects and delay spoilage. Tansy was frequently worn at that time in shoes to prevent malaria and other fevers; it has been shown, however, that some mosquito species including *Culex pipiens* take nectar from tansy flowers.

Organic Insecticide: Tansy can be used as in companion planting, and for biological pest control in organic gardens and sustainable agriculture. It is planted alongside potatoes to repel the Colorado potato beetle, with one study finding tansy reduced the beetle population by 60 to 100%.

In England tansy is placed on window sills to repel flies; sprigs are placed in bed linen to drive away pests, and it has been used as an ant repellent.

In the 1940s, distilled tansy oil mixed with fleabane, pennyroyal and diluted alcohol was a well known mosquito repellent; collectors were paid five cents a pound for tansy in full bloom. Research has found that tansy extracts do indeed repel mosquitoes, but not as effectively as chemical pesticide products containing diethyltoluamide (i.e. DEET.) In 2008, researchers in Sweden investigated the use of tansy to repel ticks, showing a 64–72% repellency for each oil constituent.

Culinary Uses: Tansy was formerly used as a flavoring for puddings and omelets, but is now almost unknown. The herbalist John Gerard noted that tansy was well known as "pleasant in taste", and he recommends tansy sweetmeats as "an especial thing against the gout, if every day for a certain space a reasonable quantitie thereof be eaten fasting." In Yorkshire, tansy and caraway seeds were traditionally used in biscuits served at funerals. During the Restoration, a "tansy" was a sweet omelette flavoured with tansy juice. In the BBC documentary "The Supersizers go ... Restoration", Allegra McEvedy described the flavour as "fruity, sharpness to it and then there's a sort of explosion of cool heat a bit like peppermint." However, the programme's presenter Sue Perkins experienced tansy toxicity.

According to liquor historian A. J. Baime, in the 19th century Tennessee whiskey magnate Jack Daniel enjoyed drinking his own whiskey with sugar and crushed tansy leaf.

Ethnomedical Use: For many years, tansy has been used as a medicinal herb despite its toxicity. 19th-century Irish folklore suggests bathing in a solution of tansy and salt as a cure for joint pain. A bitter tea made with tansy flowers has been used for centuries as an anthelmintic to treat parasitic worm infestations, and tansy cakes were traditionally eaten during Lent because it was believed that eating fish during Lent caused intestinal worms. Various *Tanacetum* species are used ethnomedically to treat migraine, neuralgia and rheumatism and as anthelmintics. Traditionally, tansy was often used for its emmenagogue effects to bring on menstruation or end an unwanted pregnancy, and pregnant women are advised to not use this herb. Research published in 2011 identified 3,5-dicaffeoylquinic acid (3,5-DCQA) and axillarin in tansy as antiviral compounds that are active against herpes simplex virus.

Other Uses: In England, bunches of tansy were traditionally placed at windows to keep out flies. Sprigs were placed in bedding and linen to drive away pests. Tansy has been widely used in gardens and homes in Melbourne, Australia to keep away ants. Some traditional dyers use tansy to produce a golden-yellow colour. The yellow flowers are dried for use in floral arrangements.

Tansy is also used as a companion plant, especially with cucurbits like cucumbers and squash, or with roses or various berries. It is thought to repel ants, cucumber beetles, Japanese beetles, squash bugs, and some kinds of flying insects, among others.

Dried tansy is used by some bee-keepers as fuel in a bee smoker.

Korean Mint (hyssop)

Who doesn't want to be minty fresh? Most of the various types of "mint" or mentha – spearmint, Korean mint, applemint, regular old mint – offer reported health benefits and medicinal properties. (Avoid pennyroyal, as it's poisonous.) Mint is famous for soothing headaches, fighting nausea, calming the stomach and reducing nervousness and fatigue. Korean mint, also called Indian mint and hyssop, is a fairly effective antiviral, making it useful for fighting colds and the flu. Whatever continent you're on, some type of mint is usually to be found. Eat whole, garnish food or make tea to get the all purpose health benefits.

In Korea, it is called (bangannip), and used for Korean pancake and stew, more specifically for Bosintang and Chu-eo-tang. Chu-eo-tang is a soup or stew cooked with Chinese muddy loach. It is called (pinyin: huò xiâng) in Chinese and it is one of the 50 fundamental herbs used in traditional Chinese medicine. Chemicals isolated from *Agastache rugosa* have some antibacterial properties. The extracts have shown antifungal activity in *in vitro* experiments. *Agastache rugosa* may have anti-atherogenic properties.Cultivation—-A moist situation is preferable, but mint will succeed in almost anysoil when once started into growth, though in dry, sandy soils it is sometimes difficult to grow, and should be planted in the coolest and dampest situations. Leaf mould, road scrapings, burnt ash and similar materials should, on the other hand, be used freely for lightening heavy, tenacious soils. It does best in a partially shaded position: if in a sheltered spot, it will start earlier in the spring than if exposed. Where a long or regular supply is required, it is a good plan to have at least one bed in a sunny and sheltered, and another in a shady position, where gatherings may be made both early and late.

As the plant is a perennial, spreading by means of its underground, creeping stems propagation may be easily effected by lifting the roots in February or March, dividing them - every piece showing a joint will grow - and planting again in shallow trenches, covering with 2 inches of soil. Six inches apart in the rows and 8 inches between the rows are the right distances to allow. Cuttings in summer or offsets in spring may also be utilized for increasing a stock. Cuttings may be taken at almost any time during the summer, always choosing the young shoots, these being struck on a shady border of light soil and kept moist, or a better plan, if possible, is to insert them in a frame, keeping them close and moist till rooted. Cuttings or young shoots will also strike freely in good-sized boxes in a heated greenhouse, in the early spring, and after the tops have been taken off two or three times for use, the plants may be hardened off and planted outside.

The beds are much benefited by an annual top-dressing of rich soil, applied towards the close of autumn, when all remaining stalks should be cut down to the ground. A liberal top-dressing of short, decayed manure, such as that from an old hot-bed or mushroom bed, annually, either in the spring, when it commences to grow, or better still, perhaps, after the first or second cutting, will ensure luxuriant growth. Frequent cuttings of shoots constitute a great drain on the plants, and if not properly nourished they will fail, more or less. To have really good mint, the plantation should be re-made about every three years, or failing that, it is essential that a good top-dressing of rich soil be added.

A good stock should be kept up, so that plenty may be available for *forcing*. Cultivators having a greenhouse can easily force mint into an earlier development of new growth than would be in the open garden. Forcing is very easy, the only preparation being the insertion of a quantity of good roots in a box of light soil, which should be placed in a temperature of about 60 degrees and watered freely as soon as growth starts. Cuttings may be made in two or three weeks. Forcing will generally be necessary from November to May - a succession being kept up by the introduction, at intervals of about three weeks, of an additional supply of roots, as forced roots soon decay. Often mint is so grown both upon and under the benches in greenhouses, and the demand for the young, tender stems and leaves during the winter is sufficient to make the plants pay well.

Pests and Diseases: Unfortunately, mint is susceptible to a disease which in some gardenshas completely destroyed it. This disease, which from its characteristic symptoms is known as Rust, is incurable. The fungus (*Puccinia Mentha*) which causes it develops inside the plant, and therefore cannot be reached by any purgicide, and as it is perennial, it cannot be got rid of by cutting off the latter. All that can be done is to prevent the spread of the disease by digging up all plants that show any sign of rust. The same ground should not be used again for mint for several years. Healthy stock should be obtained and planted in uninfected soil, some distance away. On account of this liability of mint to rust, it is advisable not to have it all in one bed, but to have several beds of it, placed at some distance from each other.

Harvesting: When the plants are breaking into bloom, the stalks should be cut a few inches above the root, on a dry day, after the dew has disappeared, and before the hot sun has taken any oil from the leaves, and dried for culinary use for the winter. All discoloured and insect-eaten leaves should be removed and the stems tied loosely into bunches and hung to dry on strings in the usual manner directed for 'bunched' herbs. The bunches should be nearly equal in length and uniform in size to facilitate packing, if intended for sale, and placed when dry in airtight boxes to prevent re-absorption of moisture.

The leaves may also be stripped from the stems as soon as thoroughly dry and rubbed through a fine sieve, so as to be freed from stalks as much as possible, or pounded in a mortar and thus powdered, stored in stoppered bottles or tins rendered airtight. If preparing for market and not for home use, the rubbed herbs will, of course, command a higher price than the bunched herbs, and should be put up in tins or bottles containing a quantity of uniform weight.

When mint is grown commercially on a large scale, it has been estimated to yield from 4 to 5 tons per acre, from which 15 to 20 cwt. of dry should be obtained. Average yields per acre are, however, taken when crops are at maturity, and an estimate of the first cutting crop is hard to form, and is likely to be less profitable than succeeding years, on account of initial expenses.

If Spearmint is being grown as a *medicinal* herb, for the sake of the volatile oil to be extracted from it, the shoots should be gathered in August, when just coming into flower, and taken to the distillery as soon as possible after picking, the British Pharmacopceia directing that oil of Spearmint be distilled from the fresh, flowering plant. It is estimated that 350 lb. of Spearmint yield 1 lb. of oil. If the distillery is not on the ground or only a short distance away, and the crop has to be dispatched by train, the cutting should take place late in the afternoon on a fine day, before the dew falls, so as to be sent off by a night train to arrive at their destination next morning, having travelled in the cool, otherwise the leaves are apt to heat and ferment, losing colour.

Chemical Constituents: Chemical compounds found in the plant include:

- Estragole, plant
- *p*-Anisaldehyde, plant
- 4-methoxycinnamaldehyde, shoot
- Pachypodol, leaf
- Methylchavicol (60.01-88.43%),
- d-Limonene
- Caryophyllene
- Hexadecanoic acid
- Linoleic acid
- Octahydro-7-methyl-methylene-4-(1-methylethyl)-1*H*-cyclopenta [1,3] cyclopropa[1,2]benzene

Alfalfa

Alfalfa is fodder for livestock for a reason: it's incredibly rich in minerals and health-promoting nutrients and compounds. With roots that grow 20 to 30 feet deep, alfalfa is considered the "father of all plants". (It also contains a high amount of protein for a green.) Alfalfa originally grew in the Mediterranean and Middle East but has now spread to most of Europe and the Americans. It can treat morning sickness, nausea, kidney stones, kidney pain and urinary discomfort. It is a powerful diuretic and has a bit of stimulant power, helping to energize after a bout with illness. It's a liver and bowel cleanser and long-term can help reduce cholesterol. You can purchase seeds and sprouts, but it's fine to eat the leaves straight from the earth.

Alfalfa, *Medicago sativa*, also called lucerne, is a perennial flowering plant in the pea family Fabaceae cultivated as an important forage crop in many countries around the world. The Spanish-Arabic (according to wiktionary and the DRAE) name alfalfa is widely used, particularly in North America and Australia. But in the UK, South Africa and New Zealand, the more commonly used name is lucerne. It superficially resembles clover, with clusters of small purple flowers followed by fruits spiralled in 2 to 3 turns containing 10-20 seeds. Alfalfa is native to a warmer temperate climate such as that of Iran (where it is thought to have originated). It has been cultivated as livestock fodder since at least the era of the ancient Greeks and Romans.

Ecology: Alfalfa is a perennial forage legume which normally lives four to eight years, but can live more than 20 years, depending on variety and climate. The plant grows to a height of up to 1 m (3 ft), and has a deep root system, sometimes stretching more than 15 m (49 ft). This makes it very resilient, especially to droughts. It has a tetraploid genome.

Alfalfa is a small-seeded crop, and has a slowly growing seedling, but after several months of establishment, forms a tough 'crown' at the top of the root system. This crown contains many shoot buds that enables alfalfa to regrow many times after being grazed or harvested.

This plant exhibits autotoxicity, which means it is difficult for alfalfa seed to grow in existing stands of alfalfa. Therefore, alfalfa fields are recommended to be rotated with other species (for example, corn or wheat) before reseeding.

Culture: Alfalfa is widely grown throughout the world as forage for cattle, and is most often harvested as hay, but can also be made into silage, grazed, or fed as greenchop. Alfalfa usually has the highest feeding value of all common hay crops. It is used less frequently as pasture. When grown on soils where it is well-adapted, alfalfa is often the highest-yielding forage plant, but its primary benefit is the combination of high yield per hectare and high nutritional quality.

Its primary use is as feed for high-producing dairy cows, because of its high protein content and highly digestible fibre, and secondarily for beef cattle, horses, sheep, and goats. Alfalfa hay is the most widely used fibre source in rabbit diets. In poultry diets, dehydrated alfalfa and alfalfa leaf concentrates are used for pigmenting eggs and meat, due to their high content in carotenoids, which are efficient for colouring egg yolk and body lipids. Humans also eat alfalfa sprouts in salads and sandwiches. Dehydrated alfalfa leaf is commercially available as a dietary supplement in several forms, such as tablets, powders and tea. Alfalfa can cause bloating in livestock, so care must be taken with livestock grazing on alfalfa because of this hazard.

Like other legumes, its root nodules contain bacteria, *Sinorhizobium meliloti*, with the ability to fix nitrogen, producing a high-protein feed regardless of available nitrogen in the soil. Its nitrogen-fixing ability (which increases soil nitrogen) and its use as an animal feed greatly improve agricultural efficiency.

Alfalfa can be sown in spring or fall, and does best on well-drained soils with a neutral pH of 6.8 – 7.5. Alfalfa requires sustained levels of potassium and phosphorus to grow well. It is moderately sensitive to salt levels in both the soil and irrigation water, although it continues to be grown in the arid southwestern United States, where salinity is an emerging issue. Soils low in fertility should be fertilized with manure or a chemical fertilizer, but correction of pH is particularly important. Usually a seeding rate of 13 – 20 kg/hectare (12 – 25 lb/acre) is recommended, with differences based upon region, soil type, and seeding method. A nurse crop is sometimes used, particularly for spring plantings, to reduce weed problems and soil erosion, but can lead to competition for light, water, and nutrients.

In most climates, alfalfa is cut three to four times a year, but it can be harvested up to 12 times per year in Arizona and southern California. Total yields are typically around eight tonnes per hectare (four short tons per acre) in temperate environments, but yields have been recorded up to 20 t/ha (16 ts per acre). Yields vary with region, weather, and the crop's stage of maturity when cut. Later cuttings improve yield, but with reduced nutritional content.

Pests and Diseases: Like most plants, alfalfa can be attacked by various pests and pathogens. Diseases often have subtle symptoms which are easily misdiagnosed and can affect leaves, roots, and stems.

Some pests, such as the alfalfa weevil, aphids, armyworms, and the potato leafhopper, can reduce alfalfa yields dramatically, particularly with the second cutting when weather is warmest. Chemical controls are sometimes used to prevent this. Alfalfa is also susceptible to root rots, including *Phytophthora*, *Rhizoctonia*, and Texas root rot.

Harvesting: When alfalfa is to be used as hay, it is usually cut and baled. Loose haystacks are still used in some areas, but bales are easier for use in transportation, storage, and feed. Ideally, the first cutting should be taken at the bud stage, and the subsequent cuttings just as the field is beginning to flower, or one-tenth bloom because carbohydrates are at their highest. When using farm equipment rather than hand-harvesting, a swather cuts the alfalfa and arranges it in windrows. In areas where the alfalfa does not immediately dry out on its own, a machine known as a mower-conditioner is used to cut the hay. The mower-conditioner has a set of rollers or flails that crimp and break the stems as they pass through the mower, making the alfalfa dry faster. After the alfalfa has dried, a tractor pulling a baler collects the hay into bales. There is also a species of yellow flowered lucerne that can grow fully in 36 hours in perfect conditions.

Several types of bales are commonly used for alfalfa. For small animals and individual horses, the alfalfa is baled into small, two-string bales, commonly named by the strands of string used to wrap it. Other bale sizes are three-string, and so on up to half-ton (six-string) "square" bales – actually rectangular, and typically about 40 x 45 x 100 cm (14 x 18 x 38 in). Small square bales weigh from 25–30 kg (50–70 lb) depending on moisture, and can be easily hand separated into "flakes". Cattle ranches use large round bales, typically 1.4 to 1.8 m (4 to 6 ft) in diametre and weighing from 500 to 1,000 kg, (1000 to 2000 lbs). These bales can be placed in stable stacks or in large feeders for herds of horses, or unrolled on the ground for large herds of cattle. The bales can be loaded and stacked with a tractor using a spike, known as a bale spear, that pierces the centre of the bale, or they can be handled with a grapple (claw) on the tractor's front-end loader. A more recent innovation is large "square" bales, roughly the same proportions as the small squares, but much larger. The bale size was set so stacks would fit perfectly on a large flatbed truck. These are more common in the western United States.

When used as feed for dairy cattle, alfalfa is often made into haylage by a process known as ensiling. Rather than drying it to make dry hay, the alfalfa is chopped finely and fermented in silos, trenches, or bags, where the oxygen supply can be limited to promote fermentation. The anaerobic fermentation of alfalfa allows it to retain high nutrient levels similar to those of fresh forage, and is also more palatable to dairy cattle than dry hay. In many cases, alfalfa silage is inoculated with different strains of microorganisms to improve the fermentation quality and aerobic stability of the silage.

Catnip

The cannabis of the cat kingdom. Famous for making cats deliriously crazy, catnip has health properties that are great for humans, too. Catnip can relieve cold symptoms (helpful if you're on a camping trip and don't have access to Nyquil). It's useful in breaking a fever as it promotes sweating. Catnip also helps stop excessive bleeding and swelling when applied rather than ingested. This mint plant (yep, another one) is also reportedly helpful in treating gas, stomach aches, and migraines. Catnip can stimulate uterine contractions, so it should not be consumed by pregnant women. It grows in the Northern Hemisphere.

Nepeta cataria, commonly known as catnip, catswort, or catmint, is a species of the genus *Nepeta* in the Lamiaceae family, native to Europe and southwestern to central Asia, and is widely naturalized elsewhere. The common name catmint can also refer to the genus as a whole.

Nepeta cataria is a short lived herbaceous perennial, growing 50–100 centimetres (20–39 in) tall and wide. It resembles a typical mint family member in appearance by having the characteristic square stem that members of the Lamiaceae plant family have, but with brown-green foliage. The coarse-toothed leaves are triangular to ovate.

The small bilabiate flowers can be white and finely spotted with pale purple or pink. They are showy and fragrant. The plant blooms from late spring through autumn.

Uses: The plant terpenoid *nepetalactone* is the main chemical constituent of the essential oil of *Nepeta cataria*. Nepetalactone can be extracted from catnip by steam distillation.

Cultivation: *Nepeta cataria* is cultivated as an ornamental plant for use in gardens. It is also grown for its attractant qualities to house cats and butterflies. The plant is drought tolerant and deer resistant. It can be a repellant for certain insects, including aphids and squash bugs.

Varieties:

- *Nepeta cataria* var. *citriodora* (or *N. cataria* subsp. *citriodora*), Lemon catnip.

Attractant: The compound "iridodial" as extracted from catnip oil has been found to attract the beneficial insect known as lacewings which eat aphids and mites.

Repellent: Nepetalactone is a mosquito and fly repellent. Oil isolated from catnip by steam distillation is a repellent against insects, in particular mosquitoes, cockroaches and termites. Research suggests that, in vitro, distilled nepetalactone repels mosquitoes ten times more effectively than DEET, the active ingredient in most insect repellents, but that it is not as effective a repellent when used on the skin.

Humans: *Nepeta cataria* can be brewed to produce a herbal tea Also used as a culinary herb for many dishes.

Medicinal: Catnip has a history of medicinal use for a variety of ailments. The plant has been consumed as a tea, juice, tincture, infusion or poultice, and has also been smoked. However, its medicinal use has fallen out of favour with the development of more commonplace pharmaceutical drugs.

Catnip contains the feline attractant nepetalactone. *Nepeta cataria* (and some other species within the genus *Nepeta*) are known for their behavioural effects on the cat family, not only on domestic cats but also other species of cats. One test showed that tigers, leopards, and lynxes all reacted strongly to catnip in a manner similar to domestic cats, while lions reacted less frequently.

With domestic cats, *N. cataria* is used as a recreational substance for pet cats' enjoyment, and catnip and catnip-laced products designed for use with domesticated cats are available to consumers. The common behaviours when cats sense the bruised leaves or stems of catnip are rubbing on the plant, rolling on the ground, pawing at it, licking it, and chewing it. Consuming much of the plant is followed by drooling, sleepiness, anxiety, leaping about and purring. Some growl, meow, scratch, or bite the hand holding it. The main response period after exposure is generally between five and fifteen minutes, after which olfactory fatigue usually sets in.

The nepetalactone in catnip acts as a feline attractant after it enters the feline's nose. Cats detect it through their olfactory epithelium, not through their vomeronasal organ. At the olfactory epithelium, the nepetalactone binds to one or more olfactory receptors.

Not all cats are affected by catnip. Roughly half to two thirds of cats are affected by the plant. The phenomenon is hereditary.

Other plants that also have this effect on cats include valerian (*Valeriana officinalis*), *Acalypha indica* (root) and plants that contain actinidine. Approximately 30% of domestic house cats who do not react to catnip will react in a similar way to Tartarian honeysuckle sawdust.

Sage

Sage is an incredibly useful herb, widely considered to be perhaps the most valuable herb. It is anti-flammatory, anti-oxidant, and antifungal. Its reputation as a panacea is even represented in its scientific name, *Salvia officinalis*, derived from the Latin word, *salvere*, which means 'to be saved'." It was used as a preservative for meat before the advent of refrigeration (eminently useful: you never know when you'll be forced to hunt in the wild). Sage aids digestion, relieves cramps, reduces diarrhea, dries up phlegm, fights colds, reduces inflammation and swelling, acts as a salve for cuts and burns, and kills bacteria. Sage apparently even brings colour back to gray hair. A definite concern when lost in the woods.

Salvia officinalis (sage, also called garden sage, or common sage) is a perennial, evergreen subshrub, with woody stems, grayish leaves, and blue to purplish flowers. It is a member of the family Lamiaceae and is native to the Mediterranean region, though it has naturalized in many places throughout the world. It has a long history of medicinal and culinary use, and in modern times as an ornamental garden plant. The common name "sage" is also used for a number of related and unrelated species.

Salvia officinalis was described by Carl Linnaeus in 1753. It has been grown for centuries in the Old World for its food and healing properties, and was often described in old herbals for the many miraculous properties attributed to it. The specific epithet, *officinalis*, refers to the plant's medicinal use—the *officina* was the traditional storeroom of a monastery where herbs and medicines were stored. *S. officinalis* has been classified under many other scientific names over the years, including six different names since 1940 alone.

Description: Cultivars are quite variable in size, leaf and flower colour, and foliage pattern, with many variegated leaf types. The Old World type grows to approximately 2 ft (0.61 m) tall and wide, with lavender flowers most common, though they can also be white, pink, or purple. The plant flowers in late spring or summer. The leaves are oblong, ranging in size up to 2.5 in (6.4 cm) long by 1 in (2.5 cm) wide. Leaves are grey-green, rugose on the upper side, and nearly white underneath due to the many short soft hairs. Modern cultivars include leaves with purple, rose, cream, and yellow in many variegated combinations.

Uses: Common sage is grown in parts of Europe for distillation of an essential oil, though other species, such as *Salvia fruticosa* may also be harvested and distilled with it.

In Britain sage has for generations been listed as one of the essential herbs, along with parsley, rosemary and thyme (as in the folk song "Scarborough Fair"). It has a savoury, slightly peppery flavour. It appears in many European cuisines, notably Italian, Balkan and Middle Eastern cookery. In British and American cooking, it is traditionally served as sage and onion stuffing, an accompaniment to roast turkey or chicken at Christmas or Thanksgiving Day. Other dishes include pork casserole, Sage Derby cheese and Lincolnshire sausages. Despite the common use of traditional and available herbs in French cuisine, sage never found favour there.

In Traditional Tamil Siddha medicine , Sage(Karpooravalli) is used for respiratory ailments like asthma and alleviating nasal discharge associated with Upper respiratory infections. Sage leaves are crushed in boiling water and the fumes are inhaled.

In the traditional Austrian medicine *Salvia officinalis* herb has been used internally (as tea or directly chewed) for treatment of disorders of the respiratory tract, mouth, gastrointestinal tract, and skin.

Salvia and "sage" are derived from the Latin *salvere* (to save), referring to the healing properties long attributed to the various *Salvia* species. It has been recommended at one time or another for virtually every ailment by various herbals. Modern evidence shows possible uses as an antisweating agent, antibiotic, antifungal, astringent, antispasmodic, estrogenic, hypoglycemic, and tonic. In a double blind, randomized and placebo-controlled trial, sage was found to be effective in the management of mild to moderate Alzheimer's disease.

The strongest active constituents of sage are within its essential oil, which contains cineole, borneol, and thujone. Sage leaf contains tannic acid, oleic acid, ursonic acid, ursolic acid, cornsole, cornsolic acid, fumaric acid, chlorogenic acid, caffeic acid, niacin, nicotinamide, flavones, flavonoid glycosides, and estrogenic substances.

Investigations have taken place into using sage as a treatment for Alzheimer's disease patients. Sage leaf extract may be effective and safe in the treatment of hyperlipidemia.

Cultivars: In favourable conditions in the garden, *S. officinalis* can grow to a substantial size (1 square metre or more), but a number of cultivars are more compact. As such they are valued as small ornamental flowering shrubs, rather than for their herbal properties. Some provide low ground cover, especially in sunny dry environments. Like many herbs they can be killed by a cold wet winter, especially if the soil is not well drained. But they are easily propagated from summer cuttings, and some cultivars are produced from seeds.

Named cultivars include:

- 'Alba', a white-flowered cultivar
- 'Aurea', golden sage
- 'Berggarten', a cultivar with large leaves, which rarely blooms, extending the useful life of the leaves
- 'Extrakta', has leaves with higher oil concentrations
- 'Icterina', a cultivar with yellow-green variegated leaves
- 'Lavandulaefolia', a small leaved cultivar
- 'Purpurascens' ('Purpurea'), a purple-leafed cultivar
- 'Tricolour', a cultivar with white, yellow and green variegated leaves

Blackberries

The blackberry is an edible fruit produced by many species in the *Rubus* genus in the Rosaceae family, hybrids among these species within the *Rubus* subgenus, and hybrids between the *Rubus* and

Idaeobatus subgenera. What distinguishes the blackberry from its raspberry relatives is whether or not the torus (receptacle or stem) 'picks-with' (i.e. stays with) the fruit. When picking a blackberry fruit, the torus does stay with the fruit. With a raspberry, the torus remains on the plant, leaving a hollow core in the raspberry fruit. The term 'bramble', a word meaning any impenetrable scrub, has traditionally been applied specifically to the blackberry or its products, though in the United States it applies to all members of the *Rubus* genus. In the western US, the term caneberry is used to refer to blackberries and raspberries as a group rather than the term bramble.

The usually black fruit is not a berry in the botanical sense of the word. Botanically it is termed an aggregate fruit, composed of small drupelets. It is a widespread and well-known group of over 375 species, many of which are closely related apomictic microspecies native throughout Europe, northwestern Africa, temperate western and central Asia and North and South America.

Growth and Anatomical Description: Blackberries are perennial plants which typically bear biennial stems ("canes") from the perennial root system.

In its first year, a new stem, the primocane, grows vigorously to its full length of 3–6 m (in some cases, up to 9 m), arching or trailing along the ground and bearing large palmately compound leaves with five or seven leaflets; it does not produce any flowers. In its second year, the cane becomes a floricane and the stem does not grow longer, but the lateral buds break to produce flowering laterals (which have smaller leaves with three or five leaflets). First- and second-year shoots usually have numerous short-curved, very sharp prickles that are often erroneously called thorns. These prickles can tear through denim with ease and make the plant very difficult to navigate around. Prickle-free cultivars have been developed. Recently the University of Arkansas has developed primocane fruiting blackberries that grow and flower on first-year growth much as the primocane-fruiting (also called fall bearing or everbearing) red raspberries do.

Unmanaged mature plants form a tangle of dense arching stems, the branches rooting from the node tip on many species when they reach the ground. Vigorous and growing rapidly in woods, scrub, hillsides, and hedgerows, blackberry shrubs tolerate poor soils, readily colonizing wasteland, ditches, and vacant lots.

The flowers are produced in late spring and early summer on short racemes on the tips of the flowering laterals. Each flower is about 2–3 cm in diametre with five white or pale pink petals.

The drupelets only develop around ovules that are fertilized by the male gamete from a pollen grain. The most likely cause of undeveloped ovules is inadequate pollinator visits. Even a small change in conditions, such as a rainy day or a day too hot for bees to work after early morning, can reduce the number of bee visits to the flower, thus reducing the quality of the fruit. Incomplete drupelet development can also be a symptom of exhausted reserves in the plant's roots or infection with a virus such as Raspberry bushy dwarf virus.

In botanical terminology, the fruit is not a berry but an aggregate fruit of numerous drupelets.

Ecology: Blackberry leaves are food for certain caterpillars; some grazing mammals, especially deer, are also very fond of the leaves. Caterpillars of the concealer moth *Alabonia geoffrella* have been found feeding inside dead blackberry shoots. When mature, the berries are eaten and their seeds dispersed by several mammals, such as the red fox and the Eurasian badger, as well as by small birds.

Blackberries grow wild throughout all parts of the United Kingdom and Ireland. They are an important element in the ecology of those countries. Harvesting the berries is a popular pastime in these countries. However, it is also considered an invasive weed, sending down its strong suckering roots amongst garden hedges and shrubs. In some parts of the world, such as in Australia, Chile, New Zealand, and the Pacific Northwest of North America, some blackberry species, particularly *Rubus armeniacus* (syn. *R. procerus*, 'Himalaya') and *Rubus laciniatus* ('Evergreen'), are naturalised and considered an invasive species and a serious weed. The blackberry tends to be red during its unripe ("green") phase, leading to an old expression that "blackberries are red when they're green". In various parts of the United States, wild blackberries are sometimes called "Black-caps", a term more commonly used for black raspberries, *Rubus occidentalis*.

As there is forensic evidence from the Iron Age Haraldskær Woman that she consumed blackberries some 2500 years ago, it is reasonable to conclude that blackberries have been eaten by humans over thousands of years.

Uses:

Food: The soft fruit is popular for use in desserts, jams, seedless jelly, and sometimes wine. It is often mixed with apples for pies and crumbles. Blackberries are also used to produce candy.

Good nectar producers, blackberry shrubs bearing flowers yield a medium to dark, fruity honey.

Phytochemical Research: Blackberries contain numerous phytochemicals including polyphenols, flavonoids, anthocyanins, salicylic acid, ellagic acid, and fibre. Anthocyanins in blackberries are responsible for their rich dark colour. Phytochemical components of blackberries, salicylic acid and ellagic acid have been associated in preliminary research with toxicity to cancer cells, including breast cancer cells.

Blackberries rank highly among fruits for in vitro antioxidant strength, particularly because of their dense content of polyphenolic compounds, such as ellagic acid, tannins, ellagitannins, quercetin, gallic acid, anthocyanins, and cyanidins. One report placed blackberry at the top of more than 1000 antioxidant foods consumed in the United States.

Nutrients: Blackberries are notable for their high nutritional contents of dietary fibre, vitamin C, vitamin K, and the essential mineral manganese. Blackberries have both soluble and insoluble fibre. One cup of blackberries (144 g) has an average of 7.6 g of fibre and contains half the daily recommended dose of vitamin C. Dietary fibre is important in maintaining a healthy digestive system, as it supports regular bowel movements.

Nutrient Content of Seeds: Blackberries contain numerous large seeds that are not always preferred by consumers. The seeds contain oil rich in omega-3 (alpha-linolenic acid) and -6 fats (linoleic acid) as well as protein, dietary fibre, carotenoids, ellagitannins and ellagic acid.

Cultivation:

Commercial Cultivation: Worldwide, Mexico is the leading producer of blackberries, with nearly the entire crop being produced for export into the off-season fresh markets in North America and Europe. The Mexican market is almost entirely from the cultivar 'Tupy' (often spelled 'Tupi', but the EMBRAPA programme in Brazil from which it was released prefers the 'Tupy' spelling.). In the US, Oregon is the leading commercial blackberry producer, producing 42.6 million pounds on 6,180 acres (25.0 km^2), in 1995 and 56.1 million pounds on 7,000 acres (28 km^2) in 2009.

Numerous cultivars have been selected for commercial and amateur cultivation in Europe and the United States. Since the many species form hybrids easily, there are numerous cultivars with more than one species in their ancestry.

'Marion' (marketed as "marionberry") is an important cultivar that was selected from seedlings from a cross between 'Chehalem' and 'Olallie' (commonly called "olallieberry") berries. 'Olallie' in turn is a cross between loganberry and youngberry. 'Marion', 'Chehalem' and 'Olallie' are just three of many trailing blackberry cultivars developed by the United States Department of Agriculture Agricultural Research Service (USDA-ARS) blackberry breeding programme at Oregon State University in Corvallis, Oregon.

The most recent cultivars released from this programme are the prickle-free cultivars 'Black Diamond', 'Black Pearl', and 'Nightfall' as well as the very early-ripening 'Obsidian' and 'Metolius'. 'Black Diamond' is now the leading cultivar being planted in the Pacific Northwest. Some of the other cultivars from this programme are 'Newberry', 'Waldo', 'Siskiyou', 'Black Butte', 'Kotata', 'Pacific', and 'Cascade'.

Trailing blackberries are vigorous and crown forming, require a trellis for support, and are less cold hardy than the erect or semi-erect blackberries. In addition to the United States's Pacific Northwest, these types do well in similar climates such as the United Kingdom, New Zealand, Chile, and the Mediterranean countries.

Semi-erect, prickle-free blackberries were first developed at the John Innes Centre in Norwich, UK, and subsequently by the USDA-ARS in Beltsville, Maryland. These are crown forming and very vigorous and need a trellis for support. Cultivars include 'Black Satin' 'Chester Thornless', 'Dirksen Thornless', 'Hull Thornless', 'Loch Ness', 'Loch Tay', 'Merton Thornless', 'Smoothstem', and 'Triple Crown'. Recently, the cultivar 'Cacanska Bestrna' (also called 'Cacak Thornless') has been developed in Serbia and has been planted on many thousands of hectares there. The University of Arkansas has developed cultivars of erect blackberries. These types are less vigorous than the semi-erect types and produce new canes from root initials (therefore they spread underground like raspberries). There are prickly and prickle-free cultivars from this programme, including 'Navaho', 'Ouachita', 'Cherokee', 'Apache', 'Arapaho', and 'Kiowa'. They are also responsible for developing the primocane fruiting blackberries such as 'Prime-Jan' and 'Prime-Jim'.

In raspberries, these types are called primocane fruiting, fall fruiting, or everbearing. 'Prime-Jim' and 'Prime-Jan' were released in 2004 by the University of Arkansas and are the first cultivars of primocane fruiting blackberry. They grow much like the other erect cultivars described above, however the canes that emerge in the spring, will flower in mid-summer and fruit in late summer or fall. The fall crop has its highest quality when it ripens in cool mild climate such as in California or the Pacific Northwest.

'Illini Hardy' a semi-erect prickly cultivar introduced by the University of Illinois is cane hardy in zone 5, where traditionally blackberry production has been problematic, since canes often failed to survive the winter. Blackberry production in Mexico has expanded enormously in the past decade. While once based on the cultivar 'Brazos', an old erect blackberry cultivar developed in Texas in 1959, the Mexican industry is now dominated by the Brazilian 'Tupy' released in the 1990s. 'Tupy' has the erect blackberry 'Comanche', and a "wild Uruguayan blackberry" as parents. Since there are no native blackberries in Uruguay, the suspicion is that the widely grown 'Boysenberry' is the male parent. In order to produce these blackberries in regions of Mexico where there is no winter chilling to stimulate flower bud development, chemical defoliation and application of growth regulators are used to bring the plants into bloom.

Diseases and Pests: As a result of blackberries belonging to the same genus as raspberries, they share the same diseases including anthracnose which can cause the berry to have uneven ripening and sap flow may also be slowed. They also share the same remedies including the Bordeaux mixture, a combination of lime, water and Copper(II) sulfate. The rows between blackberry plants must be free of weeds, blackberry suckers and grasses which may lead to pests or diseases. Fruit growers are selective when planting blackberry bushes as wild blackberries may be infected and gardeners are recommended to purchase only certified disease-free plants. The spotted-wing drosophila, *Drosophila suzukii* is a serious pest of blackberries. Unlike its vinegar fly relatives which are primarily attracted to rotting or fermented fruit, *D. suzukii* attacks fresh, ripe fruit by laying eggs under the soft skin. The larvae hatch and grow in the fruit, destroying the fruit's commercial value.

Another pest is *Amphorophora rubi*, known as the Blackberry Aphid, which not only eats blackberries but raspberries as well.

Byturus tomentosus (Raspberry beetle), *Lampronia corticella* (Raspberry Moth) and *Anthonomus rubi* (Strawberry blossom weevil) are also known to infest blackberries.

Wild Quinine

Wild quinine is a potent herb that "is used as an antiperiodic, emmenagogue, kidney, lithontripic, poultice. It has traditionally been used in alternative medicine to treat debility, fatigue, respiratory infection, gastrointestinal infection, and venereal disease." Whatever the ailment, quinine is famously helpful in treating it. Only the root and flowers are edible; avoid the plant.

Parthenium integrifolium is a species of flowering plant in the aster family known by the common names wild quinine, American feverfew, and eastern feverfew. It is native to the eastern United States. This plant is a perennial herb growing 30 to 60 centimetres tall, but known to exceed one metre at times. The glandular leaves are oval to lance-shaped and variable in size. They have serrated, toothed, or lobed edges. The inflorescence is an array of several flower heads containing whitish disc flowers and 5 to 6 ray flowers. The "flowers have a pleasant but mild medicinal fragrance."

This plant grows in disturbed areas as well as prairies, woods, and hillsides. It tolerates hot and cold climates and can be used as a garden plant in many areas. The leaves of the plant contain tannins and the plant was used for medicinal and veterinary purposes by Native Americans. The Catawba people used it as a poultice to treat burns. The ashes were applied to horses with "sore backs". The roots were made into a tea to treat dysentery.

Growing Quinine Wildflowers: A member of the sunflower family, growing quinine wildflowers are found in open woods and prairies. The best growing conditions for quinine plant include fertile, well-drained soil and full sun to light shade. Plants are easily propagated by seed and are best planted in the fall or early winter. If planting in the spring, provide 4 to 6 weeks of cold and moist stratification to improve germination.

Wild Quinine Care: Once planted and established in suitable growing condition for quinine plant, it requires very little attention. There is no need to fertilize this hardy plant. Minimal water is needed as quinine develops a thick taproot and can tolerate long periods without water. There are no known pests or diseases of the wild quinine making it a great addition to a chemical-free garden. Because its leaves are rough textured and bitter tasting, bunnies and deer tend to skip over wild quinine in rain gardens and flower beds too.

Navajo Tea

Also called greenthread, Plains Tea or Coyote Plant, this plant has been used for centuries by Native Americans to quickly relieve that most brutal and irritating of infections: the UTI (urinary tract infection). Best when made into a tea or decoction.

Thelesperma megapotamicum is better known by the Navajos as either Navajo Tea or Hopi Tea. This herb has been gathered and prepared by these tribes for many centuries.

The flower on this herb stands out in a radiant yellow beam extending from its long slender stalks. This perrennial is also known as *T. gracile*.The leaves grow opposite to each other the majority being pinnately parted out in a few lobes that are narrow. The heads are small to medium in size and are bright yellow in colour. The fruit is so small and is enbedded within the flower that it may remain unnoticed unless being specifically search for.

Plant Preparation: When gathering break stem off 2-3 inches from the topsoil. This will allow a new stem and flower to grow in its place. If gathering by the roadside, pick flowers furthest from the road.

When preparing first wash thoroughly, shake off excess water. Have a string nearby, to tie bundles (I prefer the string off a bag of Blue Bird Flour). Take two stalks if long (more than 10 inches), three to four is shorter. Fold ~3 1/2 inches from the root end, continue until all is folded. Tie string around twice and knot. This tea bundle can be now boiled and enjoyed or can be dried and stored (perferably in a dark area; ex., brown bag or canister) for later use.

Medicinal Uses:

- Considered useful for the Kidneys
- Settle the stomach
- Purify the blood
- Gonorrhea (Personal Interview)
- Nervous stimulant
- Toothache remedy

Edible:

- Beverage

Other Uses:

- Dye for basketry
- Textile Dye

Active Compounds:

- Luteolin
- Luteolin-7-O Glucoside
- Marein (Ateya)

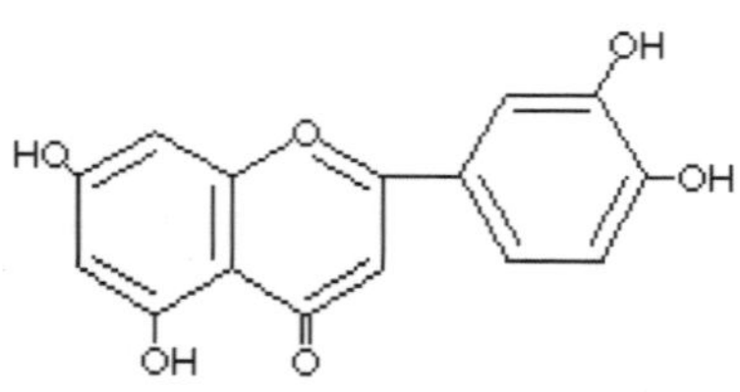

Figure: *luteolin*

Red Clover

Trifolium pratense (red clover) is a species of clover, native to Europe, Western Asia and northwest Africa, but planted and naturalised in many other regions.

It is an herbaceous, short-lived perennial plant, variable in size, growing to 20–80 cm tall. The leaves are alternate, trifoliate (with three leaflets), each leaflet 15–30 mm long and 8–15 mm broad, green with a characteristic pale crescent in the outer half of the leaf; the petiole is 1–4 cm long, with two basal stipules. The flowers are dark pink with a paler base, 12–15 mm long, produced in a dense inflorescence.

Native to Europe, Northern Africa and Western Asia, red clover is now ubiquitous worldwide. The plant's reddish pink blossoms can be used for coughs and colds, but they are an excellent detoxifier and blood cleanser as well.

Diseases: Red clover is subject to bacterial as well as fungal diseases. Other problems include parasitic nematodes (roundworms) and viruses.

Uses: It is widely grown as a fodder crop, valued for its nitrogen fixation, which increases soil fertility. For these reasons it is used as a green manure crop. Several cultivar groups have been selected for agricultural use, mostly derived from var. *sativum*. It has become naturalised in many temperate areas, including the Americas and Australasia as an escape from cultivation. Red clover is commonly used to make a sweet-tasting herbal tea. It is an ingredient in some recipes for essiac tea.

Warnings and Contraindications: In alternative medicine, red clover is promoted as a treatment for a variety of human maladies, including coughs, disorders of the lymphatic system and a variety of cancers. However, according to the American Cancer Society, "available clinical evidence does not show that red clover is effective in treating or preventing cancer, menopausal symptoms, or any other medical conditions." Dietary amounts of red clover are safe, but medicinal quantities may cause rash-like reactions, muscle ache, headache, nausea, vaginal bleeding in women, and slow blood clotting. Due to its activity on estrogen receptors, red clover is contraindicated in people with a history of breast cancer, endometriosis, ovarian cancer, uterine cancer, uterine fibroids, or other estrogen-sensitive conditions, but others have suggested the high isoflavone content counteracts this, and even provides benefits in these conditions. Due to its coumarin derivatives, it should be used in caution in individuals with coagulation disorders or currently undergoing anticoagulation therapy. It is metabolized by CYP3A4 and therefore caution should be used when taking it with other drugs using this metabolic pathway. There is evidence that red clover may cause infertility.

Sweet Marjoram

Marjoram and oregano are often used interchangeably, but the aromatic sweet marjoram is slightly different. The Greeks called it the "Joy of the Mountain" and it was revered throughout the Mediterranean for its fragrance, flavour and medicinal value. The famous French herbs de provence and Middle Eastern za'atar both use sweet marjoram. Marjoram has many uses (it's a famous digestive aid) but it is effective as an antifungal, antibacterial and disinfectant treatment in a pinch.

Cultivation: Considered a tender perennial (USDA Zones 7-9), marjoram can sometimes prove hardy even in zone 5. Marjoram is cultivated for its aromatic leaves, either green or dry, for culinary purposes; the tops are cut as the plants begin to flower and are dried slowly in the shade. It is often used in herb combinations such as *herbes de Provence* and *za'atar*. The flowering leaves and tops of marjoram are steam-distilled to produce an essential oil that is yellowish in colour (darkening to brown as it ages). It has many chemical components, some of which are borneol, camphor and pinene.

Related Species: Oregano (*Origanum vulgare*, sometimes listed with marjoram as *Origanum majorana*) is also called wild marjoram. It is a perennial common in southern Europe in dry copses and on hedge-banks, with many stout stems 30–80 cm high, bearing short-stalked, somewhat ovate leaves and clusters of purple flowers. It has a stronger flavour than marjoram. Pot marjoram or Cretan oregano (*Origanum onites*) has similar uses to marjoram.

Hardy marjoram or French marjoram, a cross of marjoram with oregano, is much more resistant to cold, but is slightly less sweet. *Origanum pulchellum* is known as showy marjoram or showy oregano.

Uses: Marjoram is used for seasoning soups, stews, dressings and sauce. *Majorana hortensis* herb has been used in the traditional Austrian medicine for treatment of disorders of the gastrointestinal tract and infections.

Burdock Herb

Burdock, or cocklebur, is a prickly, thistle-like plant that grows commonly in many parts of the world. It can get fairly big and its leaves resemble the elephant ear plant. Though the burs often get caught in pets' and livestock's fur, don't think of it only as an annoying plant. It is a highly effective treatment against poison ivy and poison oak. Plants of the genus *Arctium* have dark green leaves that can grow up to 28" (71 cm) long. They are generally large, coarse and ovate, with the lower ones being heart-shaped. They are woolly underneath. The leafstalks are generally hollow. *Arctium* species generally flower from July through to October.

The prickly heads of these plants (burrs) are noted for easily catching on to fur and clothing (being the inspiration for Velcro), thus providing an excellent mechanism for seed dispersal. Burrs cause local irritation and can possibly cause intestinal hairballs in pets. However, most animals avoid ingesting these plants. Birds are especially prone to becoming entangled with their feathers in the burrs leading to a slow death, as they are unable to free themselves.

A large number of species have been placed in genus *Arctium* at one time or another, but most of them are now classified in the related genus *Cousinia*. The precise limits between *Arctium* and *Cousinia* are hard to define; there is an exact correlation between their molecular phylogeny. The burdocks are sometimes confused with the cockleburs (genus *Xanthium*) and rhubarb (genus *Rheum*). The roots of burdock, among other plants, are eaten by the larva of the Ghost Moth (*Hepialus humuli*). The plant is used as a food plant by other Lepidoptera including Brown-tail, *Coleophora paripennella*, *Coleophora peribenanderi*, the Gothic, Lime-speck Pug and Scalloped Hazel.

The green, above-ground portions may cause contact dermatitis in humans due to the lactones the plant produces.

How to Grow Burdock:

Propagation: Burdock is grown from seed. Sow Burdock seeds in your garden as early as the ground can be worked. Cover seeds with 1/2" to 3/4" of fine garden or seed starting soil. Tamp soil. Water lightly, if the soil is dry. Seedlings germinate quickly, in about 4-7 days.

How to Grow Burdock Plants: Burdock is easy to grow. It grows readily in the wild, without day human assistance. The attention you provide to it in your garden, will promote bigger, deeper roots, and more plant and seed to harvest. Burdock likes full sun, but will tolerate partial shade. The soil should be rich, and drain well. The most important characteristic of the soil, should be loose and free of rocks and stones up to a depth of a couple of feet. This allows the roots to grow long and straight. After seedlings have grown a few inches, thin to 3" apart in rows two feet apart. Weed frequently, especially when plants are young.

Keep soil moist. Apply a general purpose fertilizer once a month.

Did you Know?: The roots of some varieties of Burdock can grow four feet long. Some plant varieties can grow up to nine feet tall.

Keep flowers and burrs picked, to promote growth of plant and root.

Tip: Fewer leaves allow denser plants and a higher yield. If you are growing Burdock just for the roots, prune some leaves as the plant grows.

Harvesting: Pick leaves and tender young stems for eating fresh. Cook like spinach. Roots take a long time to grow, over 100 days. Pick when roots are one to two feet long. Peel and eat raw or cooked. Use in soups, salads, and stir fries.

Insect and Disease: Insects can be an occasional problem. Treat with insecticide or repellents, as needed.

Uses:

Food and drink: The taproot of young burdock plants can be harvested and eaten as a root vegetable. While generally out of favour in modern European cuisine, it remains popular in Asia. In Japan, *Arctium lappa* is called "gobô"; in Korea burdock root is called "u-eong" and sold as "tong u-eong", or "whole burdock". Plants are cultivated for their slender roots, which can grow about one metre long and two centimetres across. Burdock root is very crisp and has a sweet, mild, and pungent flavour with a little muddy harshness that can be reduced by soaking julienned or shredded roots in water for five to ten minutes.

Immature flower stalks may also be harvested in late spring, before flowers appear; their taste resembles that of artichoke, to which the burdock is related. The stalks are thoroughly peeled, and either eaten raw, or boiled in salt water. Leaves are also eaten in spring in Japan when a plant is young and leaves are soft. Some *A. lappa* cultivars are specialized for this purpose. A popular Japanese dish is kinpira gobô, julienned or shredded burdock root and carrot, braised with soy sauce, sugar, mirin and/or sake, and sesame oil. Another is burdock makizushi (sushi filled with pickled burdock root; the burdock root is often artificially coloured orange to resemble a carrot).

In the second half of the 20th century, burdock achieved international recognition for its culinary use due to the increasing popularity of the macrobiotic diet, which advocates its consumption. It contains a fair amount of dietary fibre (GDF, 6g per 100g), calcium, potassium, amino acids, and is low in calories. It contains a polyphenol oxidase, which causes its darkened surface and muddy harshness by forming tannin-iron complexes. Burdock root's harshness harmonizes well with pork in miso soup (tonjiru) and with Japanese-style pilaf (takikomi gohan). Dandelion and burdock is today a soft drink that has long been popular in the United Kingdom, which has its origins in hedgerow mead commonly drunk in the mediæval period. Burdock is believed to be a galactagogue, a substance that increases lactation, but it is sometimes recommended to be avoided during pregnancy based on animal studies that show components of burdock to cause uterus stimulation. In Europe, burdock root was used as a bittering agent in beer before the widespread adoption of hops for this purpose. The American composer Christian Wolff composed a work for variable performers entitled "Burdocks" in 1970-71.

Traditional Medicine: Folk herbalists considered dried burdock to be a diuretic, diaphoretic, and a blood purifying agent. The seeds of *A. lappa* are used in traditional Chinese medicine, under the name *niubangzi*; Some dictionaries list the Chinese as just niúbàng.)

Burdock is a traditional medicinal herb that is used for many ailments. Burdock root oil extract, also called Bur oil, is currently used in Europe under the belief that it is a useful scalp treatment. Modern studies indicate that burdock root oil extract is rich in phytosterols and essential fatty acids (including rare long-chain EFAs).

Feverfew

Feverfew is a plant that has well-known and documented health properties and medicinal benefits. This anti-inflammatory can treat rheumatism, arthritis and, most famously, migraine headaches and tension headaches. It's also good for alleviating tension and general anxiety (it is a natural serotonin inhibitor). It also helps to reduce swelling and bruising. Though feverfew is most effective when taken daily, it can be a helpful pain reliever when no Advil is on hand.

Cultivation: A perennial herb, which should be planted in full sun, 38–46 cm (15–18 in) apart and grows up to 61 cm (24 in) tall. It is hardy to USDA zone 5 (–30 °C (–22 °F)) and should be cut back to the ground in the autumn. Outside of its native range it can become an invasive weed. Feverfew was native to Eurasia; specifically the Balkan Peninsula, Anatolia and the Caucasus, but cultivation has spread it around the world and it is now also found in the rest of Europe, North America and Chile.

Uses: Feverfew has been used as a herbal treatment to reduce fever and to treat headaches, arthritis and digestive problems, though scientific evidence does not support anything beyond a placebo effect.

The active ingredients in feverfew include parthenolide and tanetin. There has been some scientific interest in parthenolide, which has been shown to induce apoptosis in some cancer cell lines *in vitro* and potentially to target cancer stem cells. There are no published studies of parthenolide or feverfew in humans with cancer. The parthenolide content of commercially available feverfew supplements varies substantially, by over 40-fold, despite labelling claims of "standardization". A study found that the actual parthenolide content of these supplements bore little resemblance to the content claimed on the product label.

Long-term use of feverfew followed by abrupt discontinuation may induce a withdrawal syndrome featuring rebound headaches and muscle and joint pains. Feverfew can cause allergic reactions, including contact dermatitis. Other side effects have included gastrointestinal upset such as nausea, vomiting, abdominal pain, diarrhea, and flatulence. When the herb is chewed or taken orally it can cause mouth ulcers and swelling and numbness of the mouth. Feverfew should not be taken by pregnant women. It may interact with blood thinners and increase the risk of bleeding, and may also interact with a variety of medications metabolized by the liver.

Sweet Violet

Native to Europe and Asia, sweet violet is cultivated around the world and is a pleasant, delicate purple colour. When brewed into a syrup the plant is effective as a treatment for colds, flu and coughs or sore throat. However, when made as a tea, it is wonderfully effective for relieving headaches and muscle and body pain.

Characteristics: *V. odorata* can be distinguished by the following characteristics:

- the flowers are aromatic,
- the flowers are normally either dark violet or white,
- the leaves and flowers are all in a basal rosette,

- the style is hooked (and does not end with a rounded appendage),
- the leaf-stalks have hairs which point downwards, and
- the plant spreads with stolons (above-ground shoots).

These perennial flowers can mature at a height of 4 to 6 inches and a spread of 8 to 24 inches. The species can be found near the edges of forests or in clearings; it is also a common "uninvited guest" in shaded lawns or elsewhere in gardens.

Uses: Several cultivars have been selected for garden use, of which *V. odorata* 'Wellsiana' has gained the Royal Horticultural Society's Award of Garden Merit.

The sweet scent of this flower has proved popular throughout the generations particularly in the late Victorian period, and has consequently been used in the production of many cosmetic fragrances and perfumes. The French are also known for their violet syrup, most commonly made from an extract of violets. In the United States, this French violet syrup is used to make violet scones and marshmallows. The scent of violet flowers is distinctive with only a few other flowers having a remotely similar odour. References to violets and the desirable nature of the fragrance go back to classical sources such as Pliny and Horace when the name 'Ion' was in use to describe this flower from which the name of the distinctive chemical constituents of the flower, the ionones – is derived. In 1923 Poucher writes that the flowers are widely cultivated both in Europe and the East for their fragrance, with both the flowers and leaves being separately collected and extracted for fragrance, and flowers also collected for use in confectionary and the production of a galenical syrup. and people also make medecine from it. There is some doubt as to whether the true extract of the violet flower is still commercially available at all. It certainly was in the early 20th Century, but by the time Steffen Arctander was writing in the late 1950s and early 1960s production had "almost disappeared".

The violet leaf absolute however remains widely used in modern perfumery. The leaves are edible and contain a deal of soothing mucilage. Water extracts of the plant have some *in vitro* antibacterial properties.

Winter Savory

Winter savory is your savior against insect bites and stings. One of the most effective natural plant treatments for bug bites is originally from Europe and the Mediterranean but often shows up elsewhere thanks to global trade. In addition to being an antiseptic, it is delicious – used for flavoring meats and stews – and all parts are edible.

With so many amazing medicinal plants on the planet, be sure to look for future posts covering more. Feel free to submit your own request or share your botanical knowledge in the comments.

Cultivation and Uses: Easy to grow, it makes an attractive border plant for any culinary herb garden. It requires six hours of sun a day in soil that drains well. *S. montana 'Nana'* is a dwarf cultivar. In temperate climates it goes dormant in winter, putting out leaves on the bare stems again in the spring – do not cut the plant back, all those stems which appear dead will leaf out again. It is hardy and has a low bunching habit. Winter savory was used for hundreds of years both it and summer savory have been grown and used, virtually side by side. Both have strong spicy flavour. It goes particularly well with any type of mushroom, or in white sauces, and is very good in potato salads. Small amounts spice a regular salad well. It has a rich herbaceous aroma when crushed. It is used as a companion plant for beans, keeping bean weevils away, and also roses, reducing mildew and aphids.

Culinary Uses: In cooking, winter savory has a reputation for going very well with both beans and meats, very often lighter meats such as chicken or turkey, and can be used in stuffing. It has a strong flavour while uncooked but loses much of its flavour under prolonged cooking.

Medicinal Uses: Winter savory has been purported to have antiseptic, aromatic, carminative, and digestive benefits. It has also been used as an expectorant and in the treatment of stings. The plant has a stronger action than the closely related summer savory. Taken internally, it is said to be a remedy for colic and a cure for flatulence, whilst it is also used to treat gastro-enteritis, cystitis, nausea, diarrhoea, bronchial congestion, sore throat and menstrual disorders. It should not be prescribed for pregnant women. A sprig of the plant, rubbed onto bee or wasp stings, brings instant relief. Therapeutic-grade oil has been determined to inhibit growth of Candida albicans. The plant is harvested in the summer when in flower and can be used fresh or dried. The essential oil forms an ingredient in lotions for the scalp in cases of incipient baldness. An ointment made from the plant is used externally to relieve arthritic joints.

In traditional herbal medicine, summer savory was believed to be an aphrodisiac, while winter savory was believed to inhibit sexual desire.

4

Aromatic Plants

One of the most striking characteristics of our native floa is their fragrance.

"Aromatic" is defined as having a distinctive smell. "Aromatic" plants are given this title due a distinct fragrance or odor. Chemically, the organic compounds that cause the molecules to be aromatic form flat rings which bond easily to the receptors in the nose.

One of humans' most alert senses is the sense of smell. Fragrances can transport people through time and space, connecting them with a common bond. A single smell can remind thousands of people of the same memory or thought; the smell of grandmother's kitchen, the aroma of a sensuous relaxing bath, or a first date at the pizza parlor.

Aromatic plants have been used for thousands of years for many different purposes ranging from medicinal to religious; from protection against spirits to culinary delights.

Lemon Grass (*Cymbopogon citratus*)

Cymbopogon citratus, commonly known as lemon grass or oil grass, is a tropical plant from Southeast Asia.

Culinary Uses

Cymbopogon citratus is abundant in the Philippines and Indonesia where it is known as *tanglad* or *sereh*. Its fragrant leaves are traditionally used in cooking, particularly for *lechon* and roasted chicken.

Medicinal Uses

Cymbopogon citratus is often sold in stem form. While it can be grown in warmer temperate regions (such as the UK), it is not hardy to frost.

In the folk medicine of Brazil, it is believed to have anxiolytic, hypnotic, and anticonvulsant properties, but at least one study has found no effect on humans.

Laboratory studies have shown cytoprotective, antioxidant, and anti-inflammatory properties *in vitro*, as well as antifungal properties (though *Cymbopogon martinii* was found to be more effective in that study).

Citronellol, which is thought to possess antihypertensive properties, is an essential oil constituent from *Cymbopogon citratus*, *Cymbopogon winterianus*, and *Lippia alba*. Citronellol has been shown to lower blood pressure in rats by a direct effect on the vascular smooth muscle leading to vasodilation. In a small, randomised, controlled trial, an infusion made from *C. citratus* was used as an inexpensive remedy for the treatment of oral thrush in HIV/AIDS patients.

Lemon grass contains 65-85% citral and active ingredients such as myrcene, an antibacterial and pain reliever; citronella; citronellol; and geranilol. Hydrosteam distillation, condensation, and cooling were used to separate the oil from the water. Hydrosol or Hydrolat, as a by-product of the distillation process, is a pure natural water or plant water essence used for the production of skin care products such as lotions, creams, and facial cleansing toner in its pure form. The main products are organic unadulterated lemon grass oil (for industrial users), and "negros oil" (mixture of lemon grass oil with virgin coconut oil) used in aromatherapy.

Vanilla (*Vanilla planifolia*)

Vanilla is a flavour derived from orchids of the genus *Vanilla*, primarily from the Mexican species, flat-leaved vanilla (*V. planifolia*). The word *vanilla*, derived from the diminutive of the Spanish word *vaina* (*vaina* itself meaning sheath or pod), simply translates as little pod. Pre-Columbian Mesoamerican people cultivated the vine of the vanilla orchid, called *tlilxochitl* by the Aztecs, and Spanish conquistador Hernán Cortés is credited with introducing both vanilla and chocolate to Europe in the 1520s.

Initial attempts to cultivate vanilla outside Mexico and Central America proved futile because of the symbiotic relationship between the vanilla orchid and its natural pollinator, the local species of *Melipona* bee. Pollination is required to set the fruit from which the flavoring is derived. In 1837, Belgian botanist Charles François Antoine Morren discovered this fact and pioneered a method of artificially pollinating the plant. The method proved financially unworkable and was not deployed commercially. In 1841, Edmond Albius, a slave who lived on the French island of Réunion in the Indian Ocean, discovered at the age of 12 that the plant could be hand-pollinated. Hand-pollination allowed global cultivation of the plant.

Three major species of vanilla currently are grown globally, all of which derive from a species originally found in Mesoamerica, including parts of modern-day Mexico. The various subspecies are *Vanilla planifolia* (syn. *V. fragrans*), grown on Madagascar, Réunion, and other tropical areas along the Indian Ocean; *V. tahitensis*, grown in the South Pacific; and *V. pompona*, found in the West Indies, and Central and South America. The majority of the world's vanilla is the *V. planifolia* species, more commonly known as Bourbon vanilla (after the former name of Réunion, Île Bourbon) or Madagascar vanilla, which is produced in Madagascar and neighbouring islands in the southwestern Indian Ocean, and in Indonesia. *Leptotes bicolour* is used in the same way in South America.

Vanilla is the second most expensive spice after saffron, because growing the vanilla seed pods is labour-intensive. Despite the expense, vanilla is highly valued for its flavour, which author Frederic Rosengarten, Jr. described in *The Book of Spices* as "pure, spicy, and delicate"; he called its complex floral aroma a "peculiar bouquet". As a result, vanilla is widely used in both commercial and domestic baking, perfume manufacture and aromatherapy.

Biology

The main species harvested for vanilla is *Vanilla planifolia*. Although it is native to Mexico, it is now widely grown throughout the tropics. Indonesia and Madagascar are the world's largest producers. Additional sources include *Vanilla pompona* and *Vanilla tahitiensis* (grown in Niue and Tahiti), although the vanillin content of these species is much less than *Vanilla planifolia*.

Vanilla grows as a vine, climbing up an existing tree (also called a tutor), pole, or other support. It can be grown in a wood (on trees), in a plantation (on trees or poles), or in a "shader", in increasing orders of productivity. Its growth environment is referred to as its *terroir*, and includes not only the adjacent plants, but also the climate, geography, and local geology. Left alone, it will grow as high as possible on the support, with few flowers. Every year, growers fold the higher parts of the plant downward so the plant stays at heights accessible by a standing human. This also greatly stimulates flowering.

The distinctively flavored compounds are found in the fruit, which results from the pollination of the flower. One flower produces one fruit. *V. planifolia* flowers are hermaphroditic: They carry both male (anther) and female (stigma) organs; however, to avoid self-pollination, a membrane separates those organs. The flowers can be naturally pollinated only by bees of the *Melipona* genus found in Mexico (*abeja de monte* or mountain bee). This bee provided Mexico with a 300-year-long monopoly on vanilla production, from the time it was first discovered by Europeans and the French first transplanted the vines to their overseas colonies, until a substitute was found for them. The vines would grow, but would not fruit outside of Mexico. Growers tried to bring this bee into other growing locales, to no avail. The only way to produce fruits without the bees is artificial pollination. And today, even in Mexico, hand pollination is used extensively.

In 1836, botanist Charles François Antoine Morren was drinking coffee on a patio in Papantla (in Veracruz, Mexico) and noticed black bees flying around the vanilla flowers next to his table. He watched their actions closely as they would land and work their way under a flap inside the flower, transferring pollen in the process. Within hours, the flowers closed and several days later, Morren noticed vanilla pods beginning to form. Morren immediately began experimenting with hand pollination. A few years later in 1841, a simple and efficient artificial hand-pollination method was developed by a 12-year-old slave named Edmond Albius on Réunion, a method still used today. Using a beveled sliver of bamboo, an agricultural worker lifts the membrane separating the anther and the stigma, then, using the thumb, transfers the pollinia from the anther to the stigma. The flower, self-pollinated, will then produce a fruit. The vanilla flower lasts about one day, sometimes less, so growers have to inspect their plantations every day for open flowers, a labour-intensive task.

The fruit, a seed capsule, if left on the plant, will ripen and open at the end; as it dries, the phenolic compounds crystallize, giving the fruits a diamond-dusted appearance, which the French call *givre* (hoarfrost). It will then release the distinctive vanilla smell. The fruit contains tiny, flavorless seeds. In dishes prepared with whole natural vanilla, these seeds are recognizable as black specks.

Like other orchids' seeds, vanilla seeds will not germinate without the presence of certain mycorrhizal fungi. Instead, growers reproduce the plant by cutting: they remove sections of the vine with six or more leaf nodes, a root opposite each leaf. The two lower leaves are removed, and this area is buried in loose soil at the base of a support. The remaining upper roots will cling to the support, and often grow down into the soil. Growth is rapid under good conditions.

Cultivars

- Bourbon vanilla or Bourbon-Madagascar vanilla, produced from *V. planifolia* plants introduced from the Americas, is the term used for vanilla from Indian Ocean islands such as Madagascar, the Comoros, and Réunion, formerly the *Île Bourbon*. It is also used to describe the distinctive vanilla flavour derived from *V. planifolia* grown successfully in tropical countries such as India.
- Mexican vanilla, made from the native *V. planifolia*, is produced in much less quantity and marketed as the vanilla from the land of its origin. Vanilla sold in tourist markets around Mexico is sometimes not actual vanilla extract, but is mixed with an extract of the tonka bean, which contains coumarin. Tonka bean extract smells and tastes like vanilla, but coumarin has been shown to cause liver damage in lab animals and is banned in food in the US by the Food and Drug Administration since 1954.
- Tahitian vanilla is the name for vanilla from French Polynesia, made with the *V. tahitiensis* strain. Genetic analysis shows this species is possibly a cultivar from a hybrid-cross of *V. planifolia* and *V. odorata*. The species was introduced by French Admiral François Alphonse Hamelin to French Polynesia from the Philippines, where it was introduced from Guatemala by the Manila Galleon trade.
- West Indian vanilla is made from the *V. pompona* strain grown in the Caribbean and Central and South America.

The term French vanilla is often used to designate preparations with a strong vanilla aroma, containing vanilla grains and sometimes also containing eggs (especially egg yolks). The appellation originates from the French style of making vanilla ice cream with a custard base, using vanilla pods, cream, and egg yolks. Inclusion of vanilla varietals from any of the former French dependencies or overseas France noted for their exports may in fact be a part of the flavoring, though it may often be coincidental. Alternatively, French vanilla is taken to refer to a vanilla-custard flavour. Syrup labelled as French vanilla may include custard, caramel or butterscotch flavors in addition to vanilla.

Chemistry

There are many compounds present in the extracts of vanilla. The chemical compound vanillin (4-hydroxy-3-methoxybenzaldehyde) is primarily responsible for the characteristic flavour and smell of vanilla. Another minor component of vanilla essential oil is piperonal (heliotropin). Piperonal and other substances affect the odor of natural vanilla. Vanillin was first isolated from vanilla pods by Gobley in 1858. By 1874, it had been obtained from glycosides of pine tree sap, temporarily causing a depression in the natural vanilla industry.

Vanilla essence comes in two forms. Real seedpod extract is an extremely complicated mixture of several hundred different compounds, including acetaldehyde, acetic acid, furfural, hexanoic acid, 4-hydroxybenzaldehyde, eugenol, methyl cinnamate, and isobutyric acid. Synthetic essence consists of a solution of synthetic vanillin in ethanol. Vanillin can be easily synthesized from various raw materials, but the majority of food grade (>99% pure) vanillin is made from guaiacol.

Production

General guidelines: In general, good vanilla will only come from good vines and through careful production methods. Commercial vanilla production can be performed under open field and "greenhouse" operations. Both production systems share the following similarities:

- Plant height and number of years before producing the first grains
- Shade necessities
- Amount of organic matter needed
- A tree or frame to grow around (bamboo, coconut or *Erythrina lanceolata*)
- Labour intensity (pollination and harvest activities)

Vanilla grows best in a hot, humid climate from sea level to an elevation of 1500 m. The ideal climate has moderate rainfall, 1500–3000 mm, evenly distributed through 10 months of the year. Optimum temperatures for cultivation are 15–30 °C (59–86 °F) during the day and 15–20 °C (59–68 °F) during the night. Ideal humidity is around 80%, and under normal greenhouse conditions, it can be achieved by an evaporative cooler. However, since greenhouse vanilla is grown near the equator and under polymer (HDPE) netting (shading of 50%), this humidity can be achieved by the environment. Most successful vanilla growing and processing is done in the region within 10 to 20° of the equator.

Soils for vanilla cultivation should be loose, with high organic matter content and loamy texture. They must be well drained, and a slight slope helps in this condition. Soil pH has not been well documented, but some researchers have indicated an optimum soil pH of around 5.3. Mulch is very important for proper growth of the vine, and a considerable portion of mulch should be placed in the base of the vine. Fertilization varies with soil conditions, but general recommendations are: 40 to 60 g of N, 20 to 30 g of P_2O_5 and 60 to 100 g of K_2O should be applied to each plant per year besides organic manures, such as vermicompost, oil cakes, poultry manure and wood ash. Foliar applications are also good for vanilla, and a solution of 1% NPK (17:17:17) can be sprayed on the plant once a month. Vanilla requires organic matter, so three or four applications of mulch a year are adequate for the plant.

Propagation, Preparation and Type of Stock

Dissemination of vanilla can be achieved either by stem cutting or by tissue culture. For stem cutting, a progeny garden needs to be established. Recommendations for establishing this garden vary, but in general, trenches of 60 cm (24 in) in width, 45 cm (18 in) in depth and 60 cm (24 in) spacing for each plant are necessary. All plants need to grow under 50% shade, as well as the rest of the crop. Mulching the trenches with coconut husk and micro irrigation provide an ideal microclimate for vegetative growth. Cuttings between 60 and 120 cm (24 and 47 in) should be selected for planting in the field or greenhouse. Cuttings below 60 to 120 cm (24 to 47 in) need to be rooted and raised in a separate nursery before planting. Planting material should always come from unflowered portions of the vine. Wilting of the cuttings before planting provides better conditions for root initiation and establishment.

Before planting the cuttings, trees to support the vine must be planted at least three months before sowing the cuttings. Pits of 30 x 30 x 30 cm are dug 30 cm (12 in) away from the tree and filled with farm yard manure (vermicompost), sand and top soil mixed well. An average of 2000 cuttings can be planted per hectare (2.5 acres). One important consideration is that when planting the cuttings from the base, four leaves should be pruned and the pruned basal point must be pressed into the soil in a way such that the nodes are in close contact with the soil, and are placed at a depth of 15 to 20 cm (5.9 to 7.9 in). The top portion of the cutting is tied to the tree using natural fibres such as banana or hemp.

Tissue Culture

Tissue culture was first used as a means of creating vanilla plants during the 1980s at Tamil Nadu University. This was the part of the first project to grow *V. planifolia* in India. At that time, a shortage of vanilla planting stock was occurring in India. The approach was inspired by the work going on to tissue culture other flowering plants. Several methods have been proposed for vanilla tissue culture, but all of them begin from axillary buds of the vanilla vine. In vitro multiplication has also been achieved through culture of callus masses, protocorns, root tips and stem nodes. Description of any of these processes can be obtained from the references listed before, but all of them are successful in generation of new vanilla plants that first need to be grown up to a height of at least 30 cm (12 in) before they can be planted in the field or greenhouse.

Scheduling Considerations

In the tropics, the ideal time for planting vanilla is from September to November, when the weather is neither too rainy nor too dry, but this recommendation varies with growing conditions. Cuttings take one to eight weeks to establish roots, and show initial signs of growth from one of the leaf axils. A thick mulch of leaves should be provided immediately after planting as an additional source of organic matter. Three years are required for cuttings to grow enough to produce flowers and subsequent pods. As with most orchids, the blossoms grow along stems branching from the main vine. The buds, growing along the 6 to 10 in (15 to 25 cm) stems, bloom and mature in sequence, each at a different interval.

Pollination

Flowering normally occurs every spring, and without pollination, the blossom wilts and falls, and no vanilla bean can grow. Each flower must be hand-pollinated within 12 hours of opening. In the wild, very few natural pollinators exist, with most pollination being carried out by bees of the genus *Melipona*. These pollinators do not exist outside the orchid's home range, and even within that range, vanilla orchids have only a 1% chance of successful pollination. As a result, all vanilla grown today is pollinated by hand. A small splinter of wood or a grass stem is used to lift the rostellum or move the flap upward, so the overhanging anther can be pressed against the stigma and self-pollinate the vine. Generally, one flower per raceme opens per day, so the raceme may be in flower for over 20 days. A healthy vine should produce about 50 to 100 beans per year, but growers are careful to pollinate only five or six flowers from the 20 on each raceme. The first flowers that open per vine should be pollinated, so the beans are similar in age. These agronomic practices facilitate harvest and increases bean quality. It takes the fruits five to six weeks to develop, but it takes around six months for the bean to mature. Over-pollination will result in diseases and inferior bean quality. A vine remains productive between 12 and 14 years.

Pest and Disease Management

Most diseases come from the uncharacteristic growing conditions of vanilla. Therefore, conditions such as excess water, insufficient drainage, heavy mulch, overpollination and too much shade favour disease development. Vanilla is susceptible to many fungal and viral diseases. *Fusarium*, *Sclerotium*, *Phytopthora*, and *Colletrotrichum* species cause rots of root, stem, leaf, bean and shoot apex. These diseases can be controlled by spraying Bordeaux mixture (1%), carbendazim (0.2%) and copper oxychloride (0.2%).

Biological control of the spread of such diseases can be managed by applying to the soil *Trichoderma* (0.5 kg (1.1 lb) per plant in the rhizosphere) and foliar application of pseudomonads (0.2%). Mosaic virus, leaf curl and cymbidium mosaic potex virus are the common viral diseases. These diseases are transmitted through the sap, so affected plants must be destroyed. The insect pests of vanilla include beetles and weevils that attack the flower, caterpillars, snakes and slugs that damage the tender parts of shoot, flower buds and immature fruit, and grasshoppers that affect cutting shoot tips. If organic agriculture is practiced, insecticides are avoided, and mechanical measures are adopted for pest management. Most of these practices are implemented under greenhouse cultivation, since such field conditions are very difficult to achieve.

Artificial Vanilla

Most artificial vanilla products contain vanillin, which can be produced synthetically from lignin, a natural polymer found in wood. Most synthetic vanillin is a byproduct from the pulp used in papermaking, in which the lignin is broken down using sulfites or sulfates. However, vanillin is only one of 171 identified aromatic components of real vanilla fruits. The orchid species *Leptotes bicolour* is used as a natural vanilla replacement in Paraguay and southern Brazil.

Nonplant Vanilla Flavoring

In the United States, castoreum, the exudate from the castor sacs of mature beavers, has been approved by the Food and Drug Administration (FDA) as a food additive, often referenced simply as a "natural flavoring" in the product's list of ingredients. It is commonly used in both food and beverages, especially as vanilla and raspberry flavoring. It is also used to flavour some cigarettes and in perfume-making.

Stages of Production

Harvest: The vanilla fruit grows quickly on the vine, but is not ready for harvest until maturity—approximately six months. Harvesting vanilla fruits is as labour-intensive as pollinating the blossoms. Immature dark green pods are not harvested. Pale yellow discolouration that commences at the distal end of the fruits is an indication of the maturity of pods. Each fruit ripens at its own time, requiring a daily harvest. To ensure the finest flavour from every fruit, each individual pod must be picked by hand just as it begins to split on the end. Overmatured fruits are likely to split, causing a reduction in market value. Its commercial value is fixed based on the length and appearance of the pod.

If the fruit is more than 15 cm (5.9 in) in length, it belongs to first-quality product. The largest fruits greater than 16 cm and up to as much as 21 cm are usually reserved for the gourmet vanilla market, for sale to top chefs and restaurants. If the fruits are between 10 and 15 cm long, pods are under the second-quality category, and fruits less than 10 cm in length are under the third-quality category. Each fruit contains thousands of tiny black vanilla seeds. Vanilla fruit yield depends on the care and management given to the hanging and fruiting vines. Any practice directed to stimulate aerial root production has a direct effect on vine productivity. A five-year-old vine can produce between 1.5 and 3 kg (3.3 and 6.6 lb) pods, and this production can increase up to 6 kg (13 lb) after a few years. The harvested green fruit can be commercialized as such or cured to get a better market price.

Curing: Several methods exist in the market for curing vanilla; nevertheless, all of them consist of four basic steps: killing, sweating, slow-drying, and conditioning of the beans.

Rosemary (Rosmarinus officinalis)

Rosmarinus officinalis, commonly known as rosemary, is a woody, perennial herb with fragrant, evergreen, needle-like leaves and white, pink, purple, or blue flowers, native to the Mediterranean region. It is a member of the mint family Lamiaceae, which includes many other herbs. The name "rosemary" derives from the Latin for "dew" (*ros*) and "sea" (*marinus*), or "dew of the sea". The plant is also sometimes called anthos, from the ancient Greek word íèïò, meaning "flower".

Rosmarinus officinalis is one of 2-4 species in the genus *Rosmarinus'*. The other species most often recognized is the closely related, *Rosmarinus eriocalyx*, of the Maghreb of Africa and Iberia. The genus was named by the 18th-century naturalist and founding taxonomist Carolus Linnaeus.

Rosemary is an aromatic evergreen shrub that has leaves similar to hemlock needles. The leaves are used as a flavoring in foods such as stuffings and roast lamb, pork, chicken and turkey. It is native to the Mediterranean and Asia, but is reasonably hardy in cool climates. It can withstand droughts, surviving a severe lack of water for lengthy periods. Forms range from upright to trailing; the upright forms can reach 1.5 m (5 ft) tall, rarely 2 m (6 ft 7 in). The leaves are evergreen, 2–4 cm (0.8–1.6 in) long and 2–5 mm broad, green above, and white below, with dense, short, woolly hair. The plant flowers in spring and summer in temperate climates, but the plants can be in constant bloom in warm climates; flowers are white, pink, purple or deep blue.

Mythology

According to legend, it was draped around the Greek goddess Aphrodite when she rose from the sea, born of Uranus's semen. The Virgin Mary is said to have spread her blue cloak over a white-blossomed rosemary bush when she was resting, and the flowers turned blue. The shrub then became known as the 'Rose of Mary'.

Usage

Rosemary is used as a decorative plant in gardens and has many culinary and medical uses. The plant is said to improve the memory. The leaves are used to flavour various foods, such as stuffings and roast meats.

Cultivation

Since it is attractive and drought tolerant, rosemary is used as an ornamental plant in gardens and for xeriscape landscaping, especially in regions of Mediterranean climate. It is considered easy to grow and pest-resistant. Rosemary can grow quite large and retain attractiveness for many years, can be pruned into formal shapes and low hedges, and has been used for topiary. It is easily grown in pots. The groundcover cultivars spread widely, with a dense and durable texture.

Rosemary grows on friable loam soil with good drainage in an open, sunny position. It will not withstand waterlogging and some varieties are susceptible to frost. It grows best in neutral to alkaline conditions (pH 7–7.8) with average fertility. It can be propagated from an existing plant by clipping a shoot (from a soft new growth) 10–15 cm (4–6 in) long, stripping a few leaves from the bottom, and planting it directly into soil.

Cultivars

Numerous cultivars have been selected for garden use. The following are frequently sold:

- 'Albus' – white flowers
- 'Arp' – leaves light green, lemon-scented
- 'Aureus' – leaves speckled yellow
- 'Benenden Blue' – leaves narrow, dark green
- 'Blue Boy' – dwarf, small leaves
- 'Golden Rain' – leaves green, with yellow streaks
- 'Gold Dust' -dark green leaves, with golden streaks but stronger than 'Golden Rain'
- 'Irene' – low and lax, trailing, intense blue flowers
- 'Lockwood de Forest' – procumbent selection from 'Tuscan Blue'
- 'Ken Taylor' – shrubby
- 'Majorica Pink' – pink flowers
- 'Miss Jessop's Upright' – distinctive tall fastigiate form, with wider leaves.
- 'Pinkie' – pink flowers
- 'Prostratus' - lower groundcover
- 'Pyramidalis' (or 'Erectus') – fastigate form, pale blue flowers
- 'Remembrance' (or 'Gallipoli') - taken from the Gallipoli Peninsula
- 'Roseus' – pink flowers
- 'Salem' – pale blue flowers, cold hardy similar to 'Arp'
- 'Severn Sea' – spreading, low-growing, with arching branches, flowers deep violet
- 'Tuscan Blue' – traditional robust upright form
- 'Wilma's Gold' – yellow leaves

The following cultivars have gained the Royal Horticultural Society's Award of Garden Merit:-

- 'Miss Jessop's upright'
- 'Severn Sea'
- 'Sissinghurst blue'
- 'Benenden blue'

Culinary Use: The leaves, both fresh and dried, are used in traditional Italian cuisine. They have a bitter, astringent taste and are highly aromatic, which complements a wide variety of foods. Herbal tea can be made from the leaves. When burnt, they give off a mustard-like smell and a smell similar to burning wood, which can be used to flavour foods while barbecuing. Rosemary is high in iron, calcium and vitamin

B_6, 317 mg, 6.65 mg and 0.336 mg per 100 g, respectively. Rosemary extract has been shown to improve the shelf life and heat stability of omega 3-rich oils, which are prone to rancidity.

Fragrance: Rosemary oil is used for purposes of fragrant bodily perfumes or to emit an aroma into a room. It is also burnt as incense, and used in shampoos and cleaning products.

Traditional Medicine: Hungary water was first prepared for the Queen of Hungary Elisabeth of Poland to " ... renovate vitality of paralyzed limbs ... " and to treat gout. It was used externally and prepared by mixing fresh rosemary tops into spirits of wine. Don Quixote mixes it in his recipe of the miraculous balm of Fierabras. Rosemary has a very old reputation for improving memory and has been used as a symbol for remembrance during weddings, war commemorations and funerals in Europe and Australia. Mourners would throw it into graves as a symbol of remembrance for the dead. In Shakespeare's *Hamlet*, Ophelia says, "There's rosemary, that's for remembrance."

Rosemary contains a number of potentially biologically active compounds, including antioxidants carnosic acid and rosmarinic acid. Other chemical compounds include camphor, caffeic acid, ursolic acid, betulinic acid, rosmaridiphenol and rosmanol. Rosemary antioxidants levels are closely related to soil moisture content.

Cosmetic Use: Rosemary is used as antiseptic and anticellulite.

Folklore and Customs: In the Middle Ages, rosemary was associated with wedding ceremonies. The bride would wear a rosemary headpiece and the groom and wedding guests would all wear a sprig of rosemary, and from this association with weddings, rosemary evolved into a love charm. Newlywed couples would plant a branch of rosemary on their wedding day. If the branch grew, it was a good omen for the union and family. In 'A Modern Herbal', Mrs Grieves says, "A rosemary branch, richly gilded and tied with silken ribands of all colours, was also presented to wedding guests, as a symbol of love and loyalty." If a young person would tap another with a rosemary sprig and if the sprig contained an open flower, it was said that the couple would fall in love.

Rosemary was used as a divinatory herb. Several herbs were grown in pots and assigned the name of a potential lover. They were left to grow and the plant that grew the strongest and fastest gave the answer. Rosemary was stuffed into poppets (cloth dolls) to attract a lover or attract curative vibrations for illness. It was believed that placing a sprig of rosemary under a pillow before sleep would repel nightmares, and if placed outside the home it would repel witches. Somehow, the use of rosemary in the garden to repel witches turned into signification that the woman ruled the household in homes and gardens where rosemary grew abundantly. By the 16th century, men were known to rip up rosemary bushes to show that they, not their wives, ruled the roost. In Australia, sprigs of rosemary are worn on ANZAC Day and sometimes Remembrance Day to signify remembrance; the herb grows wild on the Gallipoli Peninsula.

Health Benefits: In research have seen to increase the activity of Quinone Reductase and to enhance the production of Nerve Growth Factor.

Oregano (Origanum vulgare)

Oregano is a common species of *Origanum*, a genus of the mint family (Lamiaceae). It is native to warm-temperate western and southwestern Eurasia and the Mediterranean region. Oregano is a perennial herb, growing from 20–80 cm tall, with opposite leaves 1–4 cm long. Oregano will grow in a pH range between 6.0 (mildly acidic) and 9.0 (strongly alkaline) with a preferred range between 6.0 and 8.0. The flowers are purple, 3–4 mm long, produced in erect spikes. It is sometimes called wild marjoram, and its close relative *O. majorana* is known as sweet marjoram.

Oregano is related to the herb marjoram, sometimes being referred to as *wild marjoram*. Oregano has purple flowers and spade-shaped, olive-green leaves. It is a perennial, although it is grown as an annual in colder climates, as it often does not survive the winter months. Oregano is planted in early spring, the plants being spaced 30 cm (12 in) apart in fairly dry soil, with full sun. It prefers a hot, relatively dry climate, but will do well in other environments.

Taxonomy

Many subspecies and strains of oregano have been developed by humans over centuries for their unique flavors or other characteristics. Tastes range from spicy or astringent to more complicated and sweet. Simple oregano sold in garden stores as *Origanum vulgare* may have a bland taste and larger, less dense leaves, and is not considered the best for culinary uses, with a taste less remarkable and pungent. It can pollinate other more sophisticated strains, but the offspring are rarely better in quality.

The related species, *Origanum onites* (Greece, Turkey) and O. syriacum (West Asia), have similar flavors. A closely related plant is marjoram from Turkey, which, however, differs significantly in taste, because phenolic compounds are missing from its essential oil. Some varieties show a flavour intermediate between oregano and marjoram.

Notable subspecies are:

- *Origanum vulgare* subsp. *gracile* (= *O. tyttanthum*) is originally from Kyrgyzstan, and has glossy green leaves and pink flowers. It grows well in pots or containers, and is more often grown for added ornamental value than other oregano. The flavour is pungent and spicy.
- *Origanum vulgare* subsp. *hirtum* (Italian oregano, Greek oregano) is a common source of cultivars with a different aroma from those of *O. v. gracile*. Growth is vigorous and very hardy, with darker green, slightly hairy foliage. Generally, it is considered the best all-purpose culinary subspecies.

Cultivars

Cultivars traded as Italian, Sicilian, etc. are usually hardy sweet marjoram (*O. ×majoricum*), a hybrid between the southern Adriatic *O. v. hirtum* and sweet marjoram (*O. majorana*). They have a reputation for sweet and spicy tones, with little bitterness, and are prized for their flavour and compatibility with various recipes and sauces.

Uses

Culinary: Oregano is an important culinary herb, used for the flavour of its leaves, which can be more flavourful when dried than fresh. It has an aromatic, warm and slightly bitter taste, which can vary in intensity. Good quality oregano may be strong enough almost to numb the tongue, but the cultivars adapted to colder climates often have a lesser flavour. Factors such as climate, seasons and soil composition may affect the aromatic oils present, and this effect may be greater than the differences between the various species of plants. Among the chemical compounds contributing to the flavour are carvacrol, thymol, limonene, pinene, ocimene, and caryophyllene.

Oregano's most prominent modern use is as the staple herb of Italian-American cuisine. Its popularity in the US began when soldiers returning from World War II brought back with them a taste for the "pizza herb", which had probably been eaten in southern Italy for centuries. There, it is most frequently used with roasted, fried or grilled vegetables, meat and fish. Unlike most Italian herbs, oregano combines well with spicy foods, which are popular in southern Italy. It is less commonly used in the north of the country, as marjoram generally is preferred.

The herb is also widely used in Turkish, Palestinian, Lebanese, Egyptian, Syrian, Greek, Portuguese, Italian, Spanish, Philippine and Latin American cuisines.

In Turkish cuisine, oregano is mostly used for flavoring meat, especially for mutton and lamb. In barbecue and kebab restaurants, it can be usually found on table, together with paprika, salt and pepper.

The dried and ground leaves are most often used in Greece to add flavour to Greek salad, and is usually added to the lemon-olive oil sauce that accompanies many fish or meat barbecues and some casseroles. Oregano is also used by chefs in the southern Philippines to eliminate the odor of carabao or water buffalo when boiling it, while simultaneously imparting flavour.

Medicinal

Hippocrates used oregano as an antiseptic, as well as a cure for stomach and respiratory ailments. A Cretan oregano (*O. dictamnus*) is still used today in Greece as a palliative for sore throat. Oregano is high in antioxidant activity, due to a high content of phenolic acids and flavonoids. In test-tube studies, it also has shown antimicrobial activity against strains of the food-borne pathogen *Listeria monocytogenes*. In the traditional Austrian medicine *Origanum vulgare* herb has been used internally (as tea) or externally (as ointment) for treatment of disorders of the gastrointestinal tract, respiratory tract, and nervous system.

In 2005, the US Federal Trade Commission brought legal action against a firm that had claimed oil of oregano treated colds and flus, and that oil of oregano taken orally treated and relieved bacterial and viral infections and their symptoms, saying the representations were false or were not substantiated at the time the representations were made, and that they were therefore a deceptive practice and false advertisements. The final stipulation on the matter said no representation as to any health benefit could be made without "...competent and reliable scientific evidence...".

Chemical Components

The essential oil of oregano is composed primarily of monoterpenoids and monoterpenes, however the concentrations of the specific compounds vary widely depending on geographic location and other varying factors. Over 60 different compounds have been identified with the primarily ones being carvacrol and thymol ranging from 0 over 80% while lesser abundant compounds include p-cymene, ã-terpinene, caryophyllene, spathulenol, germacrene-D, β-fenchyl alcohol and ä-terpineol.

Drying of the plant material effects both quantity and distribution of volatile compounds, with methods using higher heat and longer drying times having greater negative impact. A sample of fresh whole plant material found to contain 33g/kg dry weight (3.1g/kg wet) decreased to below a third after warm air convection drying. Much higher concentrations of volatile compounds are found to be towards the end of the growing season

Essential Oil

An essential oil is a concentrated hydrophobic liquid containing volatile aroma compounds from plants. Essential oils are also known as volatile oils, ethereal oils, aetherolea, or simply as the "oil of" the plant from which they were extracted, such as oil of clove. An oil is "essential" in the sense that it carries a distinctive scent, or essence, of the plant. Essential oils do not form a distinctive category for any medical, pharmacological, or culinary purpose. They are not essential for health. Essential oils are generally extracted by distillation, often by using steam. Other processes include expression or solvent extraction. They are used in perfumes, cosmetics, soaps and other products, for flavoring food and drink, and for adding scents to incense and household cleaning products.

Essential oils have been used medicinally in history. Medical applications proposed by those who sell medicinal oils range from skin treatments to remedies for cancer and often are based solely on historical accounts of use of essential oils for these purposes. Claims for the efficacy of medical treatments, and treatment of cancers in particular, are now subject to regulation in most countries. As the use of essential oils has declined in evidence-based medicine, one must consult older textbooks for much information on their use. Modern works are less inclined to generalize; rather than refer to "essential oils" as a class at all, they prefer to discuss specific compounds, such as methyl salicylate, rather than "oil of wintergreen".

Interest in essential oils has revived in recent decades with the popularity of aromatherapy, a branch of alternative medicine that claims that essential oils and other aromatic compounds have curative effects. Oils are volatilized or diluted in a carrier oil and used in massage, diffused in the air by a nebulizer, heated over a candle flame, or burned as incense. The earliest recorded mention of the techniques and methods used to produce essential oils is believed to be that of Ibn al-Baitar (1188–1248), an Andalusian physician, pharmacist and chemist.

Some Other Aromatic Plants

Bay (Laurus nobilis): *Bay* (Laurus nobilis) or sweet bay can be used fresh or dried as a culinary herb in bouquets garnis and for drying clothes on topiary or potted specimens or hedges.

Laurus nobilis is an aromatic evergreen tree or large shrub with green, glossy leaves, native to the Mediterranean region. It is one of the plants used for bay leaf seasoning in cooking. It is known as bay laurel, sweet bay, bay tree (esp. United Kingdom), true laurel, Grecian laurel, laurel tree or simply laurel. *Laurus nobilis* figures prominently in classical Greek, Roman, and Biblical culture.

Worldwide, many other kinds of plants in diverse families are also called "bay" or "laurel", generally due to similarity of foliage or aroma to *Laurus nobilis*, and the full name is used for the California bay laurel (*Umbellularia*), also in the family Lauraceae.

Figure: *A laurel shrub*

The laurel can vary greatly in size and height, sometimes reaching 10–18 metres (33–59 ft) tall. *Laurus* is a genus of evergreen trees belonging to the Laurel family, Lauraceae. The genus includes three species, whose diagnostic key characters often overlap (Mabberley 1997).

The laurel is dioecious (unisexual), with male and female flowers on separate plants. Each flower is pale yellow-green, about 1 cm diametre, and they are borne in pairs beside a leaf. The leaves are 6–12 cm long and 2–4 cm broad, with an entire (untoothed) margin. On some leaves the margin undulates. The fruit is a small, shiny black berry-like drupe about 1 cm long that contains one seed.

A recent study found considerable genetic diversity within *L. nobilis*, and that *L. azorica* is not genetically or morphologically distinct.

Ecology

Laurus nobilis is a widespread relic of the laurel forests that originally covered much of the Mediterranean Basin when the climate of the region was more humid. With the drying of the Mediterranean during the Pliocene era, the laurel forests gradually retreated, and were replaced by the more drought-tolerant sclerophyll plant communities familiar today. Most of the last remaining laurel forests around the Mediterranean are believed to have disappeared approximately ten thousand years ago, although some remnants still persist in the mountains of southern Turkey, northern Syria, southern Spain, north-central Portugal, northern Morocco, Canary Islands and in Madeira.

Chemical Constituents

The most abundant essential oil found in laurel is cineole, also called eucalyptol. The leaves contain about 1.3% essential oils (*ol. lauri folii*), consisting of 45% eucalyptol, 12% other terpenes, 3–4% sesquiterpenes, 3% methyleugenol, and other α- and β-pinenes, phellandrene, linalool, geraniol, and terpineol.

Both essential and fatty oils are present in the fruit. The fruit is pressed and water-extracted to obtain these products. The fruit contains up to 30% fatty oils and about 1% essential oils (terpenes, sesquiterpenes, alcohols, and ketones).

Food

The plant is the source of several popular herbs and one spice used in a wide variety of recipes, particularly among Mediterranean cuisines. Most commonly, the aromatic leaves are added whole to Italian pasta sauces. However, even when cooked, whole bay leaves can be sharp and abrasive enough to damage internal organs, so they are typically removed from dishes before serving, unless used as a simple garnish. Whole bay leaves have a long shelf life of about one year, under normal temperature and humidity. Bay leaves are used almost exclusively as flavour agents during the food preparation stage;

Ground bay leaves, however, can be ingested safely and are often used in soups and stocks, as well as being a common addition to a Bloody Mary. Dried laurel berries and pressed leaf oil can both be used as robust spices, and even the wood can be burnt for strong smoke flavoring.

Traditional Medicine

Aqueous extracts of bay laurel can also be used as astringents and even as a reasonable salve for open wounds.

In massage therapy, the essential oil of bay laurel is reputed to alleviate arthritis and rheumatism, while in aromatherapy, it is used to treat earaches and high blood pressure. A traditional folk remedy for rashes caused by poison ivy, poison oak, and stinging nettle is a poultice soaked in boiled bay leaves. The chemical compound lauroside B isolated from *Laurus nobilis* is an inhibitor of human melanoma (skin cancer) cell proliferation at high concentrations.

Other Uses

Bay is widely cultivated as an ornamental plant in regions with Mediterranean or oceanic climates, and as a house plant or greenhouse plant in colder regions. It is used in topiary to create single erect stems with ball-shaped, box-shaped or twisted crowns; also for low hedges. Together with a gold form, *L. nobilis* 'Aurea', it has gained the Royal Horticultural Society's Award of Garden Merit. Laurel oil is a main ingredient, and the distinguishing characteristic of Aleppo soap.

Juniper

Juniper in all its varieties has a spicy, pungent astringent smell reminiscent of gin and tonic. Use the cones and young foliage in potpourri and to light fires. Junipers vary in size and shape from tall trees, 20–40 m tall, to columnar or low spreading shrubs with long trailing branches. They are evergreen with needle-like and/or scale-like leaves. They can be either monoecious or dioecious. The female seed cones are very distinctive, with fleshy, fruit-like coalescing scales which fuse together to form a "berry"-like structure, 4–27 mm long, with 1-12 unwinged, hard-shelled seeds. In some species these "berries" are red-brown or orange but in most they are blue; they are often aromatic and can be used as a spice. The seed maturation time varies between species from 6–18 months after pollination. The male cones are similar to those of other Cupressaceae, with 6-20 scales; most shed their pollen in early spring, but some species pollinate in the autumn.

Many junipers (e.g. *J. chinensis*, *J. virginiana*) have two types of leaves: seedlings and some twigs of older trees have needle-like leaves 5–25 mm long; and the leaves on mature plants are (mostly) tiny (2–4 mm long), overlapping and scale-like. When juvenile foliage occurs on mature plants, it is most often found on shaded shoots, with adult foliage in full sunlight. Leaves on fast-growing 'whip' shoots are often intermediate between juvenile and adult.

In some species (e. g. *J. communis*, *J. squamata*), all the foliage is of the juvenile needle-like type, with no scale leaves. In some of these (e.g. *J. communis*), the needles are jointed at the base, in others (e.g. *J. squamata*), the needles merge smoothly with the stem, not jointed.

The needle-leaves of junipers are hard and sharp, making the juvenile foliage very prickly to handle. This can be a valuable identification feature in seedlings, as the otherwise very similar juvenile foliage of cypresses (*Cupressus, Chamaecyparis*) and other related genera is soft and not prickly. Juniper is the exclusive food plant of the larvae of some Lepidoptera species including *Bucculatrix inusitata* and Juniper Carpet, and is also eaten by the larvae of other Lepidoptera species such as *Chionodes electella, Chionodes viduella,* Juniper Pug and Pine Beauty; those of the tortrix moth *C. duplicana* feed on the bark around injuries or canker.

Cultivation and Uses

Juniper berries are a spice used in a wide variety of culinary dishes and best known for the primary flavoring in gin (and responsible for gin's name, which is a shortening of the Dutch word for juniper, *genever*). Juniper berries are also used as the primary flavour in the liquor Jenever and sahti-style of beers. Juniper berry sauce is often a popular flavoring choice for quail, pheasant, veal, rabbit, venison and other meat dishes. Many of the earliest prehistoric people lived in or near juniper forests which furnished them food, fuel, and wood for shelter or utensils. Many species, such as *J. chinensis* (Chinese Juniper) from eastern Asia, are extensively used in landscaping and horticulture, and as one of the most popular species for use in bonsai. It is also a symbol of longevity, strength, athleticism, and fertility.

Some junipers are susceptible to *Gymnosporangium* rust disease, and can be a serious problem for those people growing apple trees, the alternate host of the disease.

Some junipers are given the common name "cedar," including *Juniperus virginiana*, the "red cedar" that is used widely in cedar drawers. "Eastern redcedar" is the correct name for Juniperus virginiana. The lack of space between the words "red" and "cedar" indicate that this species is not a true cedar, Cedrus.

In Morocco, the tar (gitran) of the arar tree (Juniperus phoenicea) is applied in dotted patterns on bisque drinking cups. Gitran makes the water more fragrant and is said to be good for the teeth.

American Indians, such as the Navajo, have traditionally used juniper to treat diabetes. Animal studies have shown that treatment with juniper may retard the development of streptozotocin-induced diabetes in mice. Native Americans also used juniper berries as a female contraceptive. The 17th Century herbalist physician Nicholas Culpeper recommended the ripened berries for conditions such as asthma and sciatica, as well as to speed childbirth.

Juniper berries are steam distilled to produce an essential oil that may vary from colourless to yellow or pale green. Some of its chemical components are alpha pinene, cadinene, camphene and terpineol. Leaves and twigs of Juniperus virginiana are steam distilled to produce oil of juniper. Middle Tennessee and adjacent northern Alabama and southern Kentucky are the centres for this activity. The U.S. Forest Service has provided plans for the apparatus required. This work is typically done during periods of cold weather to reduce the loss of essential oil to evaporation, which is greater in warmer weather, and to take advantage of a time of year when labour might be more readily available.

Eucalyptus

Eucalyptus or gum tree is best grown as a stooled shrub with attractive juvenile foliage in a glorious blue grey. Useful for flower arrangements and potpourri. There are more than 700 species of eucalyptus, mostly native to Australia, and a very small number are found in adjacent areas of New Guinea and Indonesia. One species, *Eucalyptus deglupta,* ranges as far north as the Philippines. Only fifteen species occur outside Australia, with just nine of these not occurring in Australia. Species of eucalyptus are cultivated widely in the tropical and temperate world, including the Americas, Europe, Africa, the Mediterranean Basin, the Middle East, China and the Indian Subcontinent, though most species do not tolerate frost.mature eucalyptus may take the form of a low shrub or a very large tree. There are three main habits and four size categories that species can be divided into.

As a generalisation "forest trees" are single-stemmed and have a crown forming a minor proportion of the whole tree height. "Woodland trees" are single-stemmed although they may branch at a short distance above ground level.

"Mallees" are multi-stemmed from ground level, usually less than 10 m (33 ft) in height, often with the crown predominantly at the ends of the branchlets and individual plants may combine to form either an open or closed formation. Many mallee trees may be so low-growing as to be considered a shrub.

Two other tree forms are notable in Western Australia and described using the native names "mallet" and "marlock". The "mallet" is a small to medium-sized tree that does not produce lignotubers and has a relatively long trunk, a steeply branching habit and often a conspicuously dense terminal crown. This is the normal habit of mature healthy specimens of *Eucalyptus occidentalis, E. astringens, E. spathulata, E. gardneri, E. dielsii, E. forrestiana, E. salubris, E. clivicola* and *E.*

ornata. The smooth bark of mallets often has a satiny sheen and may be white, cream, grey, green or copper.

The term *marlock* has been variously used; in *Forest Trees of Australia* it is defined as a small tree without lignotubers but with a shorter, lower-branching trunk than a mallet. They usually grow in more or less pure stands. Clearly recognisable examples are stands of *E. platypus*, *E. vesiculosa* and the unrelated *E. stoatei*.

The term "morrell" is somewhat obscure in origin and appears to apply to trees of the western Australian wheatbelt and goldfields which have a long, straight trunk, completely rough-barked. It is now used mainly for *E. longicornis* (Red Morrell) and *E. melanoxylon* (Black Morrell).

Tree sizes follow the convention of:

- Small — to 10 m (33 ft) in height
- Medium-sized — 10–30 m (33–98 ft)
- Tall — 30–60 m (98–197 ft)
- Very tall — over 60 m (200 ft)

Leaves

Nearly all eucalyptus are evergreen but some tropical species lose their leaves at the end of the dry season. As in other members of the myrtle family, eucalyptus leaves are covered with oil glands. The copious oils produced are an important feature of the genus. Although mature Eucalyptus trees are usually towering and fully leafed, their shade is characteristically patchy because the leaves usually hang downwards.

The leaves on a mature eucalyptus plant are commonly lanceolate, petiolate, apparently alternate and waxy or glossy green. In contrast, the leaves of seedlings are often opposite, sessile and glaucous. But there are many exceptions to this pattern. Many species such as *E. melanophloia* and *E. setosa* retain the juvenile leaf form even when the plant is reproductively mature. Some species, such as *E. macrocarpa*, *E. rhodantha* and *E. crucis*, are sought-after ornamentals due to this lifelong juvenile leaf form. A few species, such as *E. petraea*, *E. dundasii* and *E. lansdowneana*, have shiny green leaves throughout their life cycle. *E. caesia* exhibits the opposite pattern of leaf development to most eucalyptus, with shiny green leaves in the seedling stage and dull, glaucous leaves in mature crowns. The contrast between juvenile and adult leaf phases is valuable in field identification.

Four leaf phases are recognised in the development of a eucalyptus plant: the 'seedling', 'juvenile', 'intermediate' and 'adult' phases. However there is no definite transitional point between the phases. The intermediate phase, when the largest leaves are often formed, links the juvenile and adult phases.

In all except a few species, the leaves form in pairs on opposite sides of a square stem, consecutive pairs being at right angles to each other (decussate). In some narrow-leaved species, for example *E. oleosa*, the seedling leaves after the second leaf pair are often clustered in a detectable spiral arrangement about a five-sided stem. After the spiral phase, which may last from several to many nodes, the arrangement reverts to decussate by the absorption of some of the leaf-bearing faces of the stem. In those species with opposite adult foliage the leaf pairs, which have been formed opposite at the stem apex, become separated at their bases by unequal elongation of the stem to produce the apparently alternate adult leaves.

Figure: Eucalyptus leucoxylon *var. 'Rosea' showing flowers and buds with operculum present*

Flowers

The most readily recognisable characteristics of eucalyptus species are the distinctive flowers and fruit (capsules or "gumnuts"). Flowers have numerous fluffy stamens which may be white, cream, yellow, pink or red; in bud, the stamens are enclosed in a cap known as an operculum which is composed of the fused sepals or petals or both. Thus flowers have no petals, but instead decorate themselves with the many showy stamens. As the stamens expand, the operculum is forced off, splitting away from the cup-like base of the flower; this is one of the features that unites the genus. The name *Eucalyptus*, from the Greek words *eu-*, well, and *kaluptos*, cover, meaning "well-covered", describes the operculum. The woody fruits or capsules are roughly cone-shaped and have valves at the end which open to release the seeds, which are waxy, rod-shaped, about 1mm in length, and yellow-brown in colour. Most species do not flower until adult foliage starts to appear; *Eucalyptus cinerea* and *Eucalyptus perriniana* are notable exceptions.

Animal Relationships

An essential oil extracted from eucalyptus leaves contains compounds that are powerful natural disinfectants and can be toxic in large quantities. Several marsupial herbivores, notably koalas and some possums, are relatively tolerant of it. The close correlation of these oils with other more potent toxins called formylated phloroglucinol compounds (euglobals, macrocarpals and sideroxylonals) allows koalas and other marsupial species to make food choices based on the smell of the leaves. For koalas, these compounds are the most important factor in leaf choice. Eucalyptus flowers produce a great abundance of nectar, providing food for many pollinators including insects, birds, bats and possums. Although eucalyptus trees are seemingly well-defended from herbivores by the oils and phenolic compounds, they have insect pests. These include the eucalyptus longhorn borer *Phoracantha semipunctata* and the aphid-like psyllids known as "bell lerps", both of which have become established as pests throughout the world wherever eucalypts are cultivated.

The eusocial beetle *Austroplatypus incompertus* makes and defends its galleries exclusively inside *Eucalyptus* plants.

Eucalyptus Oil

Eucalyptus oil is readily steam distilled from the leaves and can be used for cleaning and as an industrial solvent, as an antiseptic, for deodorising, and in very small quantities in food supplements, especially sweets, cough drops, toothpaste and decongestants. It has insect repellent properties, and is an active ingredient in some commercial mosquito repellents. *Eucalyptus globulus* is the principal source of eucalyptus oil worldwide

Lavender

Lavender pinnata has aromatic grey-green fern-like foliage with masses of bright blue flowers and a more lemony, spicier smell than more common spiky lavenders. Hyssop has a bitter, minty smell and lovely blue flowers. It makes a good low hedge to surround herb beds when clipped. Lavender (Lavandula angustifolia) has a fresh, sweet, floral, herbaceous aroma that is soothing and refreshing. Because it is the most versatile of all essential oils, no home should be without it. Lavender is an adaptogen, and therefore can assist the body when adapting to stress or imbalances. It is a great aid for relaxing and winding down before bedtime, yet has balancing properties that can also boost stamina and energy. Therapeutic-grade lavender is highly regarded for skin and beauty. It may be used to soothe and cleanse common cuts, bruises, and skin irritations. The French scientist René Gattefossé was among the first to discover these properties when he was severely burned in a laboratory explosion. Lavender may also be used to enhance the flavour of foods. Lavender has an approximate ORAC of 3,669 (TE/L). TE/L is expressed as micromole Trolox equivalent per litre.

Cultivation

The most common form in cultivation is the common or English lavender *Lavandula angustifolia* (formerly named *L. officinalis*). A wide range of cultivars can be found. Other commonly grown ornamental species are *L. stoechas*, *L. dentata*, and *L. multifida* (Egyptian lavender).

Because the cultivated forms are planted in gardens worldwide, they are occasionally found growing wild as garden escapes, well beyond their natural range. Commonly such adventitious establishment is apparently harmless at best, but in some cases *Lavandula* species have become invasive; for example, in Australia *Lavandula stoechas* has become a cause for concern; it occurs widely throughout the continent, and has been declared a noxious weed in Victoria since 1920. It also is regarded as a weed in parts of Spain.

Lavenders flourish best in dry, well-drained, sandy or gravelly soils in full sun. All types need little or no fertilizer and good air circulation. In areas of high humidity, root rot due to fungus infection can be a problem. Organic mulches can trap moisture around the plants' bases, encouraging root rot. Gravelly materials such as crushed rocks give better results.

Lavender Oil

Commercially the plant is grown mainly for the production of essential oil of lavender. This has antiseptic and anti-inflammatory properties. These extracts are also used as fragrances for bath products.

English lavender (*Lavandula angustifolia*) yields an essential oil with sweet overtones, and can be used in balms, salves, perfumes, cosmetics, and topical applications. Lavandin, *Lavandula* × *intermedia* (also known as Dutch lavender), yields a similar essential oil, but with higher levels of terpenes including camphor, which add a sharper overtone to the fragrance.

The lavandins *Lavandula* × *intermedia* are a class of hybrids of *L. angustifolia* and *L. latifolia*. The lavandins are widely cultivated for commercial use, since their flowers tend to be bigger than those of English lavender and the plants tend to be easier to harvest, but lavandin oil is regarded by some to be of a lower quality than that of English lavender, with a perfume less sweet.

Medical Uses

The essential oil was used in hospitals during World War I. Lavender is used extensively with herbs and aromatherapy. Infusions are believed to soothe insect bites, burns, and headaches. Bunches of lavender repel insects. In pillows, lavender seeds and flowers aid sleep and relaxation. An infusion of flowerheads added to a cup of boiling water is used to soothe and relax at bedtime. Lavender oil (or extract of Lavender) is used to treat acne when diluted 1:10 with water, rosewater, or witch hazel; it also treats skin burns and inflammatory conditions.

A study published in 2010 investigated anxiolytic effects and influence on sleep quality. Lavender oil with a high percentage of linalool and linalyl acetate, in the form of capsules, was generally well tolerated. It showed meaningful efficacy in alleviating anxiety and related sleep disturbances.

Lavender oil is approved for use as an anxiolytic in Germany under the name *Lasea*. A survey paper on lavender and the nervous system published in 2013 states that, "there is growing evidence suggesting that lavender oil may be an effective medicament in treatment of several neurological disorders."

Lavender may be very effective with wounds; however, Lavender Honey (created from bees feeding on lavender plants), instead of lavender essential oil has the best effects of uninfected wounds.

Health Precautions

The U.S. National Institutes of Health (NIH) states that lavender is considered likely safe in food amounts and possibly safe in medicinal amounts. NIH does not recommend the use of lavender while pregnant or breast-feeding because of lack of knowledge of its effects. It recommends caution if young boys use lavender oil because of possible hormonal effects leading to gynecomastia, and states that lavender can cause skin irritation.

Gynecomastia

A study was published in 2007 on the use of lavender and gynecomastia in prepubescent boys. Three young boys developed gynecomastia while using products containing lavender. The boys stopped using the products, and the gynecomastia went away. The researchers found that lavender and tea tree oil can cause estrogenic and antiandrogenic activities in cell cultures, and the paper states that "repeated topical exposure to lavender and tea tree oils probably caused prepubertal gynecomastia in these boys." After the study was published *The New England Journal of Medicine* received letters disputing the paper's findings, and the authors were allowed to respond.

The Aromatherapy Trade Council of the UK and the Australian Tea Tree Association have published rebuttals.

The Aromatherapy Trade Council's rebuttal states among other things that:

- It isn't likely that there would be sufficient lavender in the personal care products used in the three cases to cause an endocrine effect.
- The gyncomastia might have come from another source. The rebuttal states that although endocrine disruptors such as medications, oral contraceptives, marijuana and soy products were considered other disruptors such as organochloride pesticides, PCBs, polychlorinated dioxins, alkyl phenols, pthalates, and parabens were not. The paper states that it is not clear that the lavender and/or tea tree oil caused the gynecomastia but say that, "other components in these products may also possess endocrine-disrupting activity that contributed to the gynecomastia, but those components were not tested because we chose to evaluate only the component that was found in all the products used by the patients (lavender oil) and a chemically similar component that was found in some of the products (tea tree oil)."
- The rebuttal points out that lavender is in widespread use saying, "considering that some 200 tonnes per annum are produced of both lavender and tea tree oil, that most of this goes into personal care products, and that very little of the evidence presented for these 3 cases is convincing, the press reports of caution are premature."

The Australian Tea Tree Industry Association rebuttal states that various other plants have estrogen inducing effects in essential oil form, including, "soy, hops, garbanzo beans, red clover, lentils, flaxseed, sunflower seeds, alfalfa sprouts, liquorice, and ginseng."

A study published in 2010 from the Research Institute for Fragrance Materials did not find estrogenic activity from lavender oil in laboratory animals. A newer version of the paper became available in 2013.

Skin Irritant

In 2004 a study was published which found that, "lavender oil is cytotoxic to human skin cells in vitro". On the other hand, aromatherapist Robert Tisserand points out in a blog post that: "Any type of in vitro test is only suggestive of a possible effect. You can never assume that the same effect will take place in the living body."

A review published in 2005 on lavender essential oil states that, "Lavender is traditionally regarded as a 'safe' oil and, although it was recently reported that lavender oil, and its major constituent linalyl acetate, are toxic to human skin cells in vitro, contact dermatitis to lavender oil appears to occur at only a very low frequency."

Photosensitivity

A study was published in 2007 which looked a the relationship between various fragrances and photosensitivity. The study stated that lavender is known "to elicit cutaneous phototoxic reactions". However, the research did not find that lavender induced photohaemolysis.

Other Uses

Flower spikes are used for dried flower arrangements. The fragrant, pale purple flowers and flower buds are used in potpourris. Lavender is also used extensively as herbal filler inside sachets used to freshen linens. Dried and sealed in pouches, lavender flowers are placed among stored items of clothing to give a fresh fragrance and to deter moths. Dried lavender flowers have become recently popular for wedding confetti. Lavender is also popular in scented waters and sachets.

Citrus

The lustrous dark green leaves of most citrus plants have a zingy smell, especially grapefruit and lime, if you can avoid scale insects and mealy bugs.

Citrus is a common term and genus (*Citrus*) of flowering plants in the rue family, Rutaceae. The most recent research indicates an origin in Australia, New Caledonia and New Guinea. Some researchers believe that the origin is in the part of Southeast Asia bordered by Northeast India, Burma (Myanmar) and the Yunnan province of China, and it is in this region that some commercial species such as oranges, mandarins, and lemons originally came. Citrus fruit has been cultivated in an ever-widening area since ancient times; the best-known examples are the oranges, lemons, grapefruit, and limes.

The generic name originated in Latin, where it specifically referred to the plant now known as Citron (*C. medica*). It was derived from the ancient Greek word for cedar, êÝãñïò (*kédros*). Some believe this was because Hellenistic Jews used the fruits of *C. medica* during Sukkot (Feast of the Tabernacles) in place of a cedar cone, while others state it was due to similarities in the smell of citrus leaves and fruit with that of cedar. Collectively, *Citrus* fruits and plants are also known by the Romance loanword agrumes (literally "sour fruits").

The taxonomy and systematics of the genus are complex and the precise number of natural species is unclear, as many of the named species are hybrids clonally propagated through seeds (by apomixis), and there is genetic evidence that even some wild, true-breeding species are of hybrid origin. Cultivated *Citrus* may be derived from as few as four ancestral species. Natural and cultivated origin hybrids include commercially important fruit such as the oranges, grapefruit, lemons, some limes, and some tangerines.

Research suggests that the closely related genus *Fortunella* (kumquats), and perhaps also *Poncirus* and the Australian *Microcitrus* and *Eremocitrus*, should be included in *Citrus*; most botanists now classify *Microcitrus* and *Eremocitrus* as part of the genus *Citrus*. Two additional genera: *Triphasia* and *Clymenia* are likewise very closely related, and bear hesperidium fruits, but are not considered part of the *Citrus* genus. At least one, *Clymenia*, will hybridize with kumquats and some limes.

Description

These plants are large shrubs or small trees, reaching 5–15 m tall, with spiny shoots and alternately arranged evergreen leaves with an entire margin. The flowers are solitary or in small corymbs, each flower 2–4 cm diametre, with five (rarely four) white petals and numerous stamens; they are often very strongly scented. The fruit is a *hesperidium*, a specialised berry, globose to elongated, 4–30 cm long and 4–20 cm diametre, with a leathery rind or "peel" called a pericarp. The outermost layer of the pericarp is an "exocarp" called the flavedo, commonly referred to as the zest. The middle layer of the pericarp is the mesocarp, which in citrus fruits consists of the white, spongy "albedo", or "pith". The innermost layer of the pericarp is the endocarp. The segments are also called "liths", and the space inside each lith is a locule filled with juice vesicles, or "pulp". From the endocarp, string-like "hairs" extend into the locules, which provide nourishment to the fruit as it develops.

Citrus fruits are notable for their fragrance, partly due to flavonoids and limonoids (which in turn are terpenes) contained in the rind, and most are juice-laden. The juice contains a high quantity of citric acid giving them their characteristic sharp flavour. The genus is commercially important as many species are cultivated for their fruit, which is eaten fresh, pressed for juice, or preserved in marmalades and pickles.

They are also good sources of vitamin C and flavonoids. The flavonoids include various flavanones and flavones.

Cultivation

Citrus trees hybridise very readily – depending on the pollen source, plants grown from a Persian Lime's seeds can produce fruit similar to grapefruit. Thus all commercial citrus cultivation uses trees produced by grafting the desired fruiting cultivars onto rootstocks selected for disease resistance and hardiness.

The colour of citrus fruits only develops in climates with a (diurnal) cool winter. In tropical regions with no winter at all, citrus fruits remain green until maturity, hence the tropical "green oranges". The Persian Lime in particular is extremely sensitive to cool conditions, thus it is not usually exposed to cool enough conditions to develop a mature colour. If they are left in a cool place over winter, the fruits will change colour to yellow.

The terms "ripe" and "mature" are usually used synonymously, but they mean different things. A mature fruit is one that has completed its growth phase. Ripening is the changes that occur within the fruit after it is mature to the beginning of decay. These changes usually involve starches converting to sugars, a decrease in acids and a softening and change in the fruit's colour.

Citrus fruits are non-climacteric and respiration slowly declines and the production and release of ethylene is gradual. The fruits do not go through a ripening process in the sense that they become "tree ripe." Some fruits, for example cherries, physically mature and then continue to ripen on the tree. Other fruits, like pears, are picked when mature but before they ripen, then continue to ripen off the tree. Citrus fruits pass from immaturity to maturity to over-maturity while still on the tree. Once they are separated from the tree, they will not increase in sweetness or continue to ripen. The only way change may happen after being picked is that they will eventually start to decay.

With oranges, colour cannot be used as an indicator of ripeness because sometimes the rinds turn orange long before the oranges are ready to eat. Tasting them is the only way to know whether or not they are ready to eat.

Citrus trees are not generally frost hardy. Mandarin Oranges (*C. reticulata*) tend to be the hardiest of the common *Citrus* species and can withstand short periods down to as cold as –10 °C, but realistically temperatures not falling below –2 °C are required for successful cultivation. Tangerines, tangors and yuzu can be grown outside even in regions with more marked sub-freezing temperatures in winter, although this may affect fruit quality. A few hardy hybrids can withstand temperatures well below freezing, but do not produce quality fruit. Lemons can be commercially grown in cooler-summer/moderate-winter coastal Southern California, because sweetness is neither attained nor expected in retail lemon fruit. The related Trifoliate Orange (*Citrus trifoliata*) can survive below “20 °C; its fruit are astringent and inedible unless cooked but a few better-tasting cultivars and hybrids have been developed.

The trees thrive in a consistently sunny, humid environment with fertile soil and adequate rainfall or irrigation. Abandoned trees in valleys may suffer, yet survive, the dry summer of Central California’s Inner Coast Ranges. At any age citrus grows well enough with infrequent irrigation in partial shade, but the fruit crop is smaller. Though broadleaved, they are evergreen and do not drop leaves except when stressed. The stems of many varieties have large sharp thorns. The trees flower in the spring, and fruit is set shortly afterwards. Fruit begins to ripen in fall or early winter months, depending on cultivar, and develops increasing sweetness afterwards. Some cultivars of tangerines ripen by winter. Some, such as the grapefruit, may take up to eighteen months to ripen.

Major commercial citrus growing areas include southern China, the Mediterranean Basin (including southern Spain), South Africa, Australia, the southernmost United States, Mexico and parts of South America. In the United States, Florida, California, Arizona, and Texas are major producers, while smaller plantings are present in other Sun Belt states and in Hawaii.

As Ornamental Plants

Citrus trees grown in tubs and wintered under cover were a feature of Renaissance gardens, once glass-making technology enabled sufficient expanses of clear glass to be produced. An orangery was a feature of royal and aristocratic residences through the 17th and 18th centuries. The *Orangerie* at the Palace of the Louvre, 1617, inspired imitations that were not eclipsed until the development of the modern greenhouse in the 1840s. In the United States the earliest surviving orangery is at the Tayloe House, Mount Airy, Virginia. George Washington had an orangery at Mount Vernon.

Some modern hobbyists still grow dwarf citrus in containers or greenhouses in areas where it is too cold to grow it outdoors. Consistent climate, sufficient sunlight, and proper watering are crucial if the trees are to thrive and produce fruit. Compared to many of the usual "green shrubs", citrus trees better tolerate poor container care. For cooler winter areas, limes and lemons should not be grown, since they are more sensitive to winter cold than other citrus fruits. Hybrids with kumquats (× *Citrofortunella*) have good cold resistance.

Pests and Diseases

Citrus plants are very liable to infestation by aphids, whitefly and scale insects (e.g. California red scale). Also rather important are the viral infections to which some of these ectoparasites serve as vectors such as the aphid-transmitted *Citrus tristeza virus* which when unchecked by proper methods of control is devastating to citrine plantations. The newest threat to citrus groves in the United States is the Asian Citrus Psyllid.

The Asian citrus psyllid is an aphid-like insect that feeds on the leaves and stems of citrus trees and other citrus-like plants – but the real danger lies in that it can carry a deadly, bacterial tree disease called Huanglongbing (HLB), also known as Citrus Greening Disease.

In June 2008, the psyllid was spotted dangerously close to California – right across the international border in Tijuana, Mexico. Only a few months later, it was detected in San Diego and Imperial counties, and has since spread to Riverside, San Bernardino, Orange, Los Angeles and Ventura counties sparking quarantines in those areas. The Asian citrus psyllid has also been intercepted coming into California in packages of fruit and plants, including citrus, ornamentals, herbs and bouquets of cut flowers, shipped from other states and countries.

The foliage is also used as a food plant by the larvae of Lepidoptera (butterfly and moth) species such as the Geometridae Common Emerald (*Hemithea aestivaria*) and Double-striped Pug (*Gymnoscelis rufifasciata*), the Arctiidae Giant Leopard Moth (*Hypercompe scribonia*), *H. eridanus*, *H. icasia* and *H. indecisa*, many species in the family Papilionidae (swallowtail butterflies), and the Black-lyre Leafroller Moth (*"Cnephasia" jactatana*), a tortrix moth.

Since 2000, the Citrus Leafminer (*Phyllocnistis citrella*) has been a pest in California, boring meandering patterns through leaves. In eastern Australia, the Bronze-orange Bug (*Musgraveia sulciventris*) can be a major pest of citrus trees, particularly grapefruit. In heavy infestations it can cause flower and fruit drop and general tree stress.

European brown snails (*Helix aspersa*) can be a problem in California, though laying female Khaki Campbell and other Mallard-related ducks can be used for control.

Deficiency Diseases

Citrus plants can also develop a deficiency condition called chlorosis, characterized by yellowing leaves highlighted by contrasting leaf veins. The shriveling leaves eventually fall, and if plant loses too many, it will slowly die. This condition is often caused by an excessively high pH (alkaline soil), which prevents the plant from absorbing iron, magnesium, zinc, or other nutrients it needs to produce chlorophyll. This condition can be cured by adding an appropriate acidic fertilizer formulated for citrus, which can sometimes revive a plant to produce new leaves and even flower buds within a few weeks under optimum conditions. A soil which is too acidic can also cause problems; citrus prefers neutral soil (pH between 6 and 8). Citrus plants are also sensitive to excessive salt in the soil. Soil testing may be necessary to properly diagnose nutrient deficiency diseases.

Uses

Culinary: Many citrus fruits, such as oranges, tangerines, grapefruits, and clementines, are generally eaten fresh. They are typically peeled and can be easily split into segments. Grapefruit is more commonly halved and eaten out of the skin with a spoon. There are special spoons (grapefruit spoons) with serrated tips designed for this purpose. Orange and grapefruit juices are also very popular breakfast beverages. More acidic citrus, such as lemons and limes, are generally not eaten on their own. Meyer Lemons can be eaten 'out of hand' with the fragant skin; they are both sweet and sour. Lemonade or limeade are popular beverages prepared by diluting the juices of these fruits and adding sugar. Lemons and limes are also used as garnishes or in cooked dishes. Their juice is used as an ingredient in a variety of dishes; it can commonly be found in salad dressings and squeezed over cooked meat or vegetables.

A variety of flavours can be derived from different parts and treatments of citrus fruits. The rind and oil of the fruit is generally very bitter, especially when cooked, and so is often combined with sugar. The fruit pulp can vary from sweet and tart to extremely sour. Marmalade, a condiment derived from cooked orange and lemon, can be especially bitter, but is usually sweetened to cut the bitterness and produce a jam-like result. Lemon or lime is commonly used as a garnish for water, soft drinks, or cocktails. Citrus juices, rinds, or slices are used in a variety of mixed drinks. The colourful outer skin of some citrus fruits, known as zest, is used as a flavouring in cooking; the white inner portion of the peel, the pith, is usually avoided due to its bitterness. The zest of a citrus fruit, typically lemon or an orange, can also be soaked in water in a coffee filter, and drunk.

Medical

Oranges were historically used for their high content of vitamin C, which prevents scurvy. Scurvy is caused by vitamin C deficiency, and can be prevented by having 10 milligrams of vitamin C a day. An early sign of scurvy is fatigue.

If ignored, later symptoms are bleeding and bruising easily. British sailors were given a ration of citrus fruits on long voyages to prevent the onset of scurvy, hence the British nickname of Limey. Pectin is a structural heteropolysaccharide contained in the primary cell walls of plants. Limes and lemons as well as oranges and grapefruits are among the highest in this level.

After consumption, the peel is sometimes used as a facial cleanser. A peel of lemons or orange is commonly used as a means to moisten medical cannabis when stored with it. Before the development of fermentation-based processes, lemons were the primary commercial source of citric acid.

Citrus fruit intake is associated with a reduced risk of stomach cancer. Also, citrus fruit juices, such as orange, lime and lemon, may be useful for lowering the risk of specific types of kidney stones. Grapefruit is another fruit juice that can be used to lower blood pressure because it interferes with the metabolism of calcium channel blockers. Lemons have the highest concentration of citrate of any citrus fruit, and daily consumption of lemonade has been shown to decrease the rate of stone formation.

Orange Oil

Orange oil is an essential oil produced by cells within the rind of an orange fruit (*Citrus sinensis* fruit). In contrast to most essential oils, it is extracted as a by-product of orange juice production by centrifugation, producing a cold-pressed oil. It is composed of mostly (greater than 90%) d-limonene, and is often used in place of pure d-limonene. D-limonene can be extracted from the oil by distillation.

Limonene

Limonene gives citrus fruit their familiar aroma, and is therefore used in perfume and household cleaners for its fragrance. It is also an effective, environmentally friendly, and relatively safe solvent, which makes it an active ingredient of choice in many applications, such as, adhesive and stain removers, cleaners of various sorts, and strippers. Limonene is also highly useful in agriculture.

Composition

The compounds inside an orange oil varies with each different oil extraction. Composition variety happens as a result of regional and seasonal changes as well as the method used for extraction. Several hundred compounds have been identified with gas chromatograph-mass spectrometry. Most of the substances in the oil belong to the terpene group with limonene being the dominant one. Long chain aliphatic hydrocarbon alcohols and aldehydes like 1-octanol and octanal are second important group of substances.

Biological Pest Control

Orange oil can be used in biological pest control green pesticides. It can kill an ant, as well as a whole colony of ants. Orange oil also erases an ant's scent-pheromone trail indicators and disrupt re-infestation activities in ants. Their use in organic farming is getting increasingly important because of their non-toxic nature. They are commonly used in kitchens to safely repel insects.

Orange oil is also known to be useful to control or exterminate termites of the Drywood (Kalotermitidae) variety.

Cedar

Many coniferous trees have delicious pine-smelling foliage and cones which can be used to light fires and wood that is used to discourage moths.

Cedar, also known as *Cedrus*, is a genus of coniferous trees in the plant family *Pinaceae*, native to the mountains of Himalaya and the Mediterranean region. The oil obtained from cedarwood is one of the oldest ingredients used in perfumery. Cedarwood has a distinctive woody, spicy-resinous scent that has been used as a base note of many famous perfumes. Cedarwood is also one of the most commonly used perfume notes, mentioned at least once in the fragrant opus of every perfumer.

Cedar is an evergreen tree that grows up to 30–40 metres, and has a silvery blue or greenish needle-shaped leaves tipped with yellow flowers. The bark of a tree is reddish to brown, even though a precise description depends on the type of a tree. True cedars are native to North African and Asian mountain regions, and commonly found in Lebanon, Deodar and Atlas. Lebanon Cedar (*Cedrus libani*) is distinguished by its cones with smooth scales; Deodar Cedar (*Cedrus deodara*) has a bright green to pale green cones, while Atlas Cedar (*Cedrus atlantica*) has a distinctive a silvery gray bark and blue-green needled leaves. Although termed cedarwood oils, most of the oils belonging to this aromatic group are obtained from distilling junipers and cypresses rather than true cedars.

Red Virginia cedar (*Juniperus virginiana*), Texas cedar (*Juniperus mexicana*), and Chinese cedar (*Chamaecyparis funebris*) are the main sources of cedarwood oil used in perfumery today. This tree contains chemical components responsible for enchanting scent of woody and ambery types of fragrances. Cedarwood is very aromatic and known to be a natural repellent to moths and other insects. For this reason, it is often used for chests that store clothing and blankets, and for shoe trees that can absorb moisture and de-odorise.

Red Virginia Cedar (Juniperus virginiana)

Cedar oil is obtained from the foliage, and sometimes the wood and roots of the tree. The essence is derived in the process of steam distillation with the yield from wood chips and sawdust of somewhere about 35 %. Cedar wood oil is yellowish or brown in colour and has a characteristic, clean, balsamic woody odor with a pale undertone of sandalwood.

The oil was used as the base for paints by the ancient Sumerians and in the ancient Egyptian embalming practices. In aromatherapy, it is used to calm and balance the energies, promote spirituality, cure urinary tract infections, clear up dandruff and help to discharge phlegm. Today, we mainly use cedarwood oil for its aromatic properties, in a range of fragrance applications.

Myrtle (Myrtus communis)

Myrtle (*Myrtus communis*), with its fluffy flowers and leaves that smell of eucalyptus when crushed, needs shelter, but should be grown more frequently.

Myrtus communis, the common myrtle or true myrtle, is native across the northern Mediterranean region.

The plant is an evergreen shrub or small tree, growing to 5 metres (16 ft) tall. The leaf is entire, 3–5 cm long, with a fragrant essential oil.

The star-like flower has five petals and sepals, and numerous stamens. Petals usually are white. The flower is pollinated by insects.

The fruit is a round berry containing several seeds, most commonly blue-black in colour. A variety with yellow-amber berries is also present. The seeds are dispersed by birds that eat the berries.

Myrtus communis, the Common Myrtle, is widely cultivated as an ornamental plant for use as a shrub in gardens and parks. It is often used as a hedge plant, with its small leaves shearing cleanly.

When trimmed less frequently, it has numerous flowers in late summer. It requires a long hot summer to produce its flowers, and protection from winter frosts.

The species and the subspecies *M. communis* subsp. *tarentina* have gained the Royal Horticultural Society's Award of Garden Merit.

Cultivation Details

Succeeds in any reasonably good soil so long as it is well-drained. Prefers a moderately fertile well-drained neutral to alkaline loam in a sunny position. Succeeds in dry soils. A very ornamental plant, when fully dormant it is hardy to between -10 and -15°c, so long as it is sheltered from cold drying winds[200], though it does withstand quite considerable maritime exposure. The young growth in spring can be damaged by late frosts.

This species does not succeed outdoors in the colder parts of Britain. A moderately fast-growing plant when young but soon slowing with age. There are a number of named varieties. 'Tarentina' with narrow small leaves is hardier than the type and is especially wind-resistant, 'Microphylla' is a dwarf form and 'Leucocarpa' has white berries. Myrtle is often cultivated in the Mediterranean, where the plant is regarded as a symbol of love and peace and is much prized for use in wedding bouquets. The foliage is strongly aromatic. Any pruning is best carried out in the spring. Plants in this genus are notably resistant to honey fungus.

Propagation

Pre-soak the seed for 24 hours in warm water and then sow it in late winter in a greenhouse. Prick out the seedlings into individual pots as soon as they are large enough to handle and grow them on in the greenhouse for at least their first winter.

Plant them out into their permanent positions in late spring or early summer, after the last expected frosts. Cuttings of half-ripe wood, 7 - 10cm with a heel, July/August in a frame. Pot up in the autumn and overwinter in a cold frame. Plant out in late spring. High percentage. Cuttings of mature wood of the current seasons growth, 7 - 12cm with a heel, November in a shaded and frost free frame. Plant out in late spring or early autumn.

Culinary

Myrtus communis, the Common Myrtle, is used in the islands of Sardinia and Corsica to produce an aromatic liqueur called *Mirto* by macerating it in alcohol. *Mirto* is one of the most typical drinks of Sardinia and comes in two varieties: *mirto rosso* (red) produced by macerating the berries, and *mirto bianco* (white) produced from the less common yellow berries and sometimes the leaves.

Medicinal

Myrtle occupies a prominent place in the writings of Hippocrates, Pliny, Dioscorides, Galen, and the Arabian writers.

In several countries, particularly in Europe and China, there has been a tradition for prescribing this substance for sinus infections. A systematic review of herbal medicines used for the treatment of rhinosinusitis concluded that the evidence that any herbal medicines are beneficial in the treatment of rhinosinusitis is limited, and that for *Myrtus* there is insufficient data to verify the significance of clinical results.

Lemon Verbena

Aloysia citrodora is a species of flowering plant in the verbena family Verbenaceae, native to western South America. Common names include lemon verbena and lemon beebrush. It was brought to Europe by the Spanish and the Portuguese in the 17th century and cultivated for its oil.

Lemon verbena is a perennial shrub or subshrub growing to 2–3 m high. The 8 cm long glossy, pointed leaves are slightly rough to the touch and emit a powerful scent reminiscent of lemon when bruised (hence the Latin specific epithet *citrodora*—lemon-scented).

Sprays of tiny lilac or white flowers appear in late Spring or early Summer. It is sensitive to cold, losing leaves at temperatures below 0 °C (32 °F) although the wood is hardy to –10 °C (14 °F).

Due to its many culinary uses, it is widely listed and marketed as a plant for the herb garden.

Uses

Lemon verbena leaves are used to add a lemon flavour to fish and poultry dishes, vegetable marinades, salad dressings, jams, puddings, Greek yogurt and beverages. It also is used to make herbal teas, or added to standard tea in place of actual lemon (as is common with Moroccan tea). It can also be used to make a sorbet. In addition, it has anti-*Candida albicans* activity. In the European Union, Verbena essential oils (Lippia citriodora Kunth.) and derivatives other than absolute are prohibited when used as a fragrance ingredient.

Moderate antioxidant supplementation with lemon verbena extract protects neutrophils against oxidative damage, decreasing the signs of muscular damage in chronic running exercise without blocking the cellular adaptation to exercise.

Lippia citriodora extract shows antioxidant properties that could play an important role in modulating GSH-reductase activity in lymphocytes and erythrocytes and protecting plasma from exercise oxidative damage.

Lemon verbena extract containing 25% verbascoside showed strong antioxidant capacity, especially in a lipophilic environment, which was higher than expected as concluded from the antioxidant capacity of pure verbascoside, probably due to synergistic effects. The capacity of verbascoside to act as an effective radical scavenger in lipophilic environments was also shown.

Verbascoside-enriched extracts might have interesting applications in cosmetic, nutraceuticals or functional food. Although some "in vitro" genotoxicity of verbascoside has been reported on human lymphocytes with an involvement of PARP-1 and p53 proteins, subsequent "in vivo" tests reported no genotoxicity for high dosage oral administration.

Chemistry

The major isolates in lemon verbena oil are citral (30–35%), nerol and geraniol.

Synonyms

Synonyms for lemon verbena are *Verbena triphylla* L'Hér., *Verbena citriodora* Cav., *Lippia triphylla, Lippia citriodora, Aloysia citriodora* (Cav.) Ort.

Garden History

The first European botanist who publicly noticed this plant was the French Philibert Commerson, who collected in Buenos Aires on his botanical circumnavigation with Bougainville, about 1767. The plant had already been quietly imported directly into the Real Jardín Botánico de Madrid, where in 1797 professors Casimiro Gómez Ortega and Antonio Palau y Verdera named it, though they did not yet effectively publish it, *Aloysia citrodora* in Latin and "Hierba de la Princesa" in Spanish, to compliment Maria Louisa of Parma, Princess of Asturias the wife of the Garden's patron Infante Carlos de Borbon, Prince of Asturias and son of king Carlos III. The name was later effectively published in the first volume of Palau's *Parte Práctica de Botánica* in 1784.

Unofficial importations from Spanish America seldom fared well: when another French botanist Joseph Dombey landed his collections at Cadiz in 1785 they were impounded and left to rot in warehouses, while he was refused permission even to have seeds planted. Among the bare handful of plants Dombey had assembled during eight years at Lima, lemon verbena survived.

Meanwhile Gómez Ortega sent seeds and specimens of the plant to Charles Louis L'Héritier de Brutelle in Paris; L'Héritier published it as *Verbena triphylla* in the second fascicle his *Stirpes Novae*. published in December 1785 or January 1786. From Paris John Sibthorpe, professor of Botany at Oxford, obtained the specimen that he introduced to British horticulture: by 1797 lemon verbena was common in greenhouses around London, and its popularity as essential in a fragrant bouquet increased through the following century.

This plant has gained the Royal Horticultural Society's Award of Garden Merit.

Soil, Planting, and Care

Many gardeners choose to grow lemon verbena in a pot and bring the pots indoors for the winter, after the leaves drop.

Plant lemon verbena in loose, well-drained soil rich in organic matter. Drainage is a key to success with this herb, which will die if roots stay constantly wet. If you're growing lemon verbena in a container, choose one at least 12 inches in diametre to give roots room to spread. A larger pot also insulates roots somewhat against soil temperature changes. Burying the container in the ground provides the best insulation for roots, but it's a risky choice.

If plants root into surrounding garden soil, when you remove the container in the fall, severing the roots will likely trigger leaf drop.Full sun yields best growth and the most flavorful leaves, although plants in southernmost and desert regions benefit from light afternoon shade. In Northernmost areas, siting a plant near a white wall or fence will surround the plant with reflected light, which enhances growth. If plants receive more shade than sun, stems will be spindly and sprawling and leaves will lack strong essential oil levels. Lemon verbena is a heavy feeder and, unlike many herbs, benefits from frequent fertilization. In early spring and throughout the growing season, fertilize lemon verbena with Bonnie Herb & Vegetable Plant Food. This organically based fertilizer is low in salt and won't cause ugly brown leaf tips. During periods of active growth, fertilize plants every four weeks outdoors, every two weeks indoors.

Lemon verbena typically drops its leaves when temperatures dip below 40 degrees F, entering dormancy. It's possible to overwinter lemon verbena outdoors in Zone 8, but it's wise to help plants harden off. To do this, reduce watering a few weeks prior to the typical onset of below-freezing temperatures.

In northern zones, carry plants indoors before cold weather arrives, or wait until a cold snap causes the plant to drop leaves, then move it indoors. Most likely, the move from outside to inside will cause the plant to drop all its leaves. Many gardeners let the weather trigger leaf drop to avoid indoor clean-up and prevent carrying insects inside. Thin plants before bringing indoors, removing spindly stems. Save these stems to dry and scent dresser drawers and closets.

Avoid overwatering dormant (leafless) plants. This is a common way gardeners kill lemon verbena, whether it's growing in planting beds or pots.

Troubleshooting

Spider mites and whiteflies adore lemon verbena. Some gardeners won't grow it because they feel this herb attracts those pests. Avoid carrying these pests indoors by allowing plants to stay outdoors until leaves drop.Lemon verbena drops its leaves prior to entering dormancy in the fall and also in response to stress. Situations that trigger leaf drop include root disturbance, an intense cold draft, quick temperature change, or transplanting. Plants also seem to enter dormancy in response to shortening day length.During dormancy, don't overwater plants. New growth typically emerges eventually (in spring for overwintering plants).

Harvest and Storage

Lemon verbena responds well to regular trimming, which keeps the plant from getting too lanky. This branch is sprouting back after being cut.

Harvest leaves as needed throughout the growing season. Each time you snip a stem, new growth will emerge at the whorl of leaves beneath the cut. This growth pattern makes frequent cutting necessary to keep stems from becoming leggy.To make a large picking, cut plants back by half. Dry leaves individually on screens or bundle stems together and hang upside down in a dark, dry place. Store dried leaves in sealed containers in a dark place. To release flavour, crumble leaves finely just before using. You can freeze lemon verbena, whole or chopped, in ice cube trays filled with water. You can also blend chopped leaves into softened butter. Store butter in an airtight container in the refrigerator for a few weeks, or form into balls and freeze on a cookie sheet. Store frozen balls in zipper bags, using them to flavour vegetables and fish or spread on bread or pancakes.

5

Tuber Crops

Tubers are various types of modified plant structures that are enlarged to store nutrients. They are used by plants to survive the winter or dry months, to provide energy and nutrients for regrowth during the next growing season, and as a means of asexual reproduction. There are both *stem* and *root* tubers.

Figure: *Oca tubers*

Stem Tubers

A stem tuber forms from thickened rhizomes or stolons. The top sides of the tuber produce shoots that grow into typical stems and leaves and the under sides produce roots. They tend to form at the sides of the parent plant and are most often located near the soil surface. The below-ground stem tuber is normally a short-lived storage and regenerative organ developing from a shoot that branches off a mature plant. The offspring or new tubers, are attached to a parent tuber or form at the end of a hypogeogenous rhizome. In the autumn the plant dies except for the new offspring stem tubers which have one dominant bud, which in spring regrows a new shoot producing stems and leaves, in summer the tubers decay and new tubers begin to grow. Some plants also form smaller tubers and/or tubercules which act like seeds, producing small plants that resemble (in morphology and size) seedlings. Some stem tubers are long lived such as those of tuberous begonia but many tuberous plants have tubers that survive only until the plants have fully leafed out, at which point the tuber is reduced to a shriveled up husk.

Stem tubers generally start off as enlargements of the hypocotyl section of a seedling but also sometimes include the first node or two of the epicotyl and the upper section of the root. The stem tuber has a vertical orientation with one or a few vegetative buds on the top and fibrous roots produced on the bottom from a basal section, typically the stem tuber has an oblong rounded shape.

Tuberous begonia and Cyclamen are commonly grown stem tubers. Mignonette vine (*Anredera cordifolia*) produces aerial stem tubers on 12-to-25-foot-tall (3.7 to 7.6 m) vines, the tubers fall to the ground and grow. *Plectranthus esculentus* of the mint family Lamiaceae, produces tuberous under ground organs from the base of the stem, weighing up to 1.8 kg per tuber, forming from axillary buds producing short stolons that grow into tubers.

Tuberous Begonia

Tuberous begonias (*Begonia* × *tuberhybrida* Voss) are a group of Begonia cultivars, regarded as some of the most spectacular of the genus.

One of the first hybrids produced was *B. x sedenii* in 1870, a cross between *B. boliviensis*, collected by botanist Richard Pearce and a species from the Andes. Another species from Peru, *B.davisii* (named after Walter Davis), was also used in early breeding.

Cultivation

Tuberous begonias grow best in partial shade or filtered sunlight. Excessive exposure to sunlight can result in burnt flowers and leaves. However, too much shade will result in lush foliage and few flowers.

Tuberous Begonias are frost-tender plants that thrive where they receive bright light but little or no direct sun. Given an early start, they put on a glorious display all summer long. We grow our Begonias in pots and hanging baskets, but they can also be planted in the ground (once the danger of frost has passed). Either way, unless you live in a frost-free climate, you must lift and store the tuberous roots in fall to carry the plants through winter.

Starting Tubers in Flats or Pots

To get a jump on a short season, plant tubers on arrival in flats (shallow containers with drainage holes) or 4-5in pots. The potting mix should have a light texture and be well drained; a soilless mix, mixed 3 to 1 with builder's sand, meets both requirements. Begin by placing potting mix and sand in a plastic tub or bucket. Slowly add water and stir until the mix is moist but not soggy.

Put moistened mix in the container, stopping about 1 1/2in shy of the rim. Handling the tubers with care (especially if they have begun to produce new growth, which is very fragile), place them, hollow side up, on top of the potting mix. Space the tubers 2-4in apart in a flat (one tuber per 4-5in pot) and cover them with 1/2in of potting mix. Then water sparingly and place the container in a window that provides bright but indirect light. Keep the potting mix moist but not soggy. To hasten growth, set the container on a heating mat or a radiator (with a few magazines between container and radiator to prevent overheating). Tubers that have not already begun to sprout when you receive them will generally show signs of growth within 2-6 weeks after planting.

Transplanting

After the first two leaves have emerged, transplant tubers started in flats into 4-5in pots or, provided the danger of frost has passed, into a lightly shaded outdoor bed that has been amended with organic matter such as compost or peat moss. Lift and move the tubers carefully to avoid damaging the roots, and set the top of each tuber 1-2in below the surface of the potting mix or garden soil.

Begonias planted in the ground should be positioned so that the points of the leaves aim at the viewer, because the blooms will face the same direction. Plants in 4-5in pots should be moved into 7-10in pots when the roots fill the pots. If time and energy permit, a final move to 12in pots will yield especially spectacular plants. Set pots outdoors in a bright but not sunny location when the threat of frost has passed. Tuberous Begonias will not grow well in deep shade.

Staking

When upright varieties are 4-6in tall, push a heavy, 18-20in long bamboo stake (or one of our cushioned, steel Begonia Stakes, item 56110) into the potting mix or garden soil on the side of the plant opposite the points of the leaves. Place the stake a few inches away from the main stem to avoid injuring the tuber. Fasten the plant to the stake with garden twine or soft plastic tape looped in a figure-eight around stem and stake. As the plant grows, you may need more ties to provide additional support.

Summer Care

Tuberous Begonias thrive in soil that is evenly moist but well drained. Soggy soil can cause the stems to snap off at thebase. Fertilize plants once a month with a balanced (20-20-20), water-soluble fertilizer mixed as directed. Keep plants tidy by removing spent flowers; cut the flowers off close to the stem using a sharp knife.The only disease that may trouble your plants is powdery mildew, a fungus that appears as white powder on the leaves. Powdery mildew is easier to prevent than it is to cure, and placing your plants in a location where air can circulate freely around them is the best prevention. If you've had trouble with powdery mildew on Tuberous Begonias in the past, you can prevent future outbreaks by spraying with a mild fungicide that you can prepare yourself by mixing one tablespoon of baking soda and 2 or 3 drops of insecticidal soap in a gallon of tepid water. Spray this solution every 10 days during hot, humid weather. Once mildew appears, the only effective remedy is to spray promptly with a commercial fungicide. Follow the directions on the label carefully.

Overwintering

Allow plants to grow through November (or until frost) to store energy for the next season. Force container-grown plants into dormancy by gradually withholding water. Dig plants grown in the ground with a ball of soil and let them dry out in a shed or on the garage floor. When the stems break free from the tubers, shake off excess soil and allow the tubers to cure in the sun for about 4 days. Then store them in dry peat moss or sand in open flats in a cool (45°-50°F), dry place. Replant the tubers as suggested above in late winter.

Growing Hanging Basket Begonias

Hanging Basket Begonias—varieties with trailing stems—require much the same care as upright Begonias, except that they look their best in a shallow container that can be suspended from the eaves of a house or from an arbor. One Hanging Basket Begonia tuber in a 12in container makes for a spectacular and long-lasting display. Please note: If the stems of a Hanging Basket Begonia grow upright and refuse to trail over the edge of the container, plants are not receiving enough light.

Cyclamen

Cyclamen is a genus of 23 species of perennials growing from tubers, valued for their flowers with upswept petals and variably patterned leaves. Cyclamen species are native to Europe and the Mediterranean Basin east to Iran, with one species in Somalia.

It was traditionally classified in the family Primulaceae but recently has been reclassified in the family Myrsinaceae.

In the APG III system (2009 onwards), Myrsinaceae was not recognized, but was sunk into Primulaceae, which in that system is circumscribed very broadly, as Myrsinoidea sub-family.

Description

Cyclamens have a tuber, from which the flowers and roots grow. In most species, leaves come up in autumn, grow through the winter, and die in spring, then the plant goes dormant through the dry Mediterranean summer.

Tuber

Figure: *A tuber of Cyclamen purpurascens with three floral trunks*

The storage organ of the cyclamen is a round tuber that develops from the hypocotyl (the stem of a seedling). It is often mistakenly called a corm, but a corm (found in crocuses for example) has a papery tunic and a basal plate from which the roots grow. The storage organ of the cyclamen has no papery covering and, depending on the species, roots may grow out of any part. It is therefore properly classified as a tuber (somewhat like a potato).

The tuber may produce roots from the top, sides, or bottom, depending on the species. *Cyclamen persicum* and *Cyclamen coum* root from the bottom; *Cyclamen hederifolium* roots from the top and sides. *Cyclamen graecum* has thick anchor roots on the bottom.

The shape of the tuber may be near spherical, as in *Cyclamen coum* or flattened, as in *Cyclamen hederifolium*. In some older specimens of *Cyclamen purpurascens* and *Cyclamen rohlfsianum*, growing points on the tuber become separated by shoulders of tissue, and the tuber becomes misshapen. In most other species, the tuber is round in old age.

Leaves and flowers sprout from growing points on the top of the tuber. Growing points that have lengthened and become like woody stems are known as floral trunks.

The size of the tuber varies depending on species. In *Cyclamen hederifolium*, older tubers commonly reach 24 cm (9.4 in) across, but in *Cyclamen parviflorum*, tubers do not grow larger than 2 cm (1 in) across.

Leaves

Leaves sprout from growing points or floral trunks on top of the tuber. Each leaf grows on its own stem. Leaf stems in early growth may be distinguished from flower stems by the direction their tips curl: tips of leaf stems curl upwards, while tips of flower stems curl downwards.

The shape of the leaves varies between the species, and even between different specimens of the same species. *Cyclamen hederifolium* and *Cyclamen repandum* usually have leaves shaped like ivy, with angles and lobes, *Cyclamen coum* has nearly round leaves, and *Cyclamen persicum* has heart-shaped leaves with a pointed tip. The leaf edge may be smooth, as in *Cyclamen coum* subsp. *coum*, or finely toothed, as in *Cyclamen graecum*. The colour of the upper side of leaves is variable, even within a species. Most species have leaves variegated in several shades of green and silver, either in an irregular pattern of blotches or an arrowhead or Christmas tree shape. In cultivation, cyclamens are selected for striking or unusual leaf patterns. The colour of the lower side of leaves is often shiny, and is from plain green to rich red or purple.

Most cyclamen species originate from the Mediterranean, where summers are hot and dry and winters are cool and wet, and are summer-dormant: their leaves sprout in the autumn, remain through the winter, and wither the next spring. *Cyclamen purpurascens* and *Cyclamen colchicum*, however, originate from cooler regions in mountains, and their leaves remain through the summer and wither only after the next year's leaves have developed.

Flowers

Flowering time may be any month of the year, depending on the species. *Cyclamen hederifolium* and *Cyclamen purpurascens* bloom in summer and autumn, *Cyclamen persicum* and *coum* bloom in winter, and *Cyclamen repandum* blooms in spring.

Figure: *On left C. persicum (without auricles); on right C. hederifolium (with auricles)*

Each flower is on a stem coming from a growing point on the tuber. In all species, the stem is bent 150-180° at the tip, so that the nose of the flower faces downwards. Flowers have 5 petals, bent outwards or up, sometimes twisted, and connected at the base into a cup, and five sepals behind the cup.

Petal shape varies depending on species, and sometimes within the same species. *Cyclamen repandum* has petals much longer than wide, *Cyclamen coum* has stubby, almost round petals, and *Cyclamen hederifolium* usually has petals with proportions between the two.

Petal colour may be white, pink, or purple, often with darker colour on the nose. Many species have a pink form and a white form, but a few have only one colour, such as *Cyclamen balearicum*, which is always white.

The dark colour on the flower nose varies in shape: *Cyclamen persicum* has a smooth band, *Cyclamen hederifolium* has a streaky V, and *Cyclamen coum* has an M-shaped splotch with two white or pink "eyes" beneath.

In some species, such as *Cyclamen hederifolium*, the petal edges at the nose are curved outwards into *auricles* (Latin for "little ears"). Most species, like *Cyclamen persicum*, have no auricles.

In most species, the style protrudes 1–3 mm out of the nose of the flower, but the stamens are inside the flower. In *Cyclamen rohlfsianum*, however, the cone of anthers sticks out prominently, about 2–3 mm (0.08–0.12 in) beyond the rim of the corolla, similar to shooting-stars (*Dodecatheon*).

Fruit

The flower stem coils or bends when the fruit begins to form. The stems of *Cyclamen hederifolium* and *Cyclamen coum* coil starting at the end, *Cyclamen persicum* arches downwards but does not curl, *Cyclamen rohlfsianum* coils starting near the tuber, and *Cyclamen graecum* coils both directions, starting at the middle.

The fruit is a round pod that opens by several flaps or teeth at maturity and contains numerous sticky seeds, brown at maturity. Natural seed dispersal is by ants (myrmecochory), which eat the sticky covering and then discard the seeds.

Cultivation and Uses

Figure: *Cyclamen persicum 'Stirling', a cultivar*

Cyclamen are commonly grown for their flowers, both outdoors and indoors in pots. Several species, particularly *Cyclamen hederifolium*, are hardy and can be grown outdoors in mild climates such as northwest Europe and the Pacific Northwest of North America.

Hardiness: Cyclamen species range from frost-hardy to frost-tender.

The most frost-hardy species, such as *Cyclamen purpurascens*, *Cyclamen hederifolium*, *Cyclamen coum*, and *Cyclamen cilicium*, tolerate temperatures down to –20 °C (–4 °F). *Cyclamen hederifolium* has even survived prolonged freezing and temperatures down to –30 °C (–22 °F).

Others, such as *Cyclamen repandum*, survive temperatures down to –14 °C (7 °F), but not prolonged freezing below this temperature.

Others, such as *Cyclamen graecum*, tolerate frost as low as –4 °C (25 °F) for a few hours.

Others, such as *Cyclamen africanum*, *Cyclamen persicum*, and *Cyclamen rohlfsianum*, only tolerate mild and brief frost.

Florist's Cyclamen: The cyclamen commonly sold by florists is *C. persicum*, which is frost-tender. Selected cyclamen cultivars can have white, bright pink, red or purple flowers. While flowering, florists' cyclamens should be kept below 20 °C (68 °F), with the night time temperatures preferably between 6.5 °C to 15 °C (44 °F to 59 °F). Temperatures above 20 °C (68 °F) may induce the plant to go dormant.

Ecology

Conservation: In many areas within the native range, cyclamen populations have been severely depleted by collection from the wild, often illegally, for the horticultural trade; some species are now endangered as a result. However, in a few areas, plant conservation charities have educated local people to control the harvest carefully at a sustainable level, including sowing seed for future crops, both sustaining the wild populations and producing a reliable long-term income. Many cyclamen are also propagated in nurseries without harm to the wild plants.

Caterpillar Food Source: Cyclamen species are eaten by the caterpillars of The Gothic moth.

Evolution

Cyclamen diversity in the Mediterranean has been studied extensively to understand how the species remain distinct (Debussche et al., 2000, 2002, 2003) and how they have reacted to the dramatic climate changes in the region. Certain climate change models suggest many species could become extinct in their current range within the next 50 years.

Ginger

Ginger is the underground rhizome of the ginger plant with a firm, striated texture. The flesh of the ginger rhizome can be yellow, white or red in colour, depending upon the variety. It is covered with a brownish skin that may either be thick or thin, depending upon whether the plant was harvested when it was mature or young.

The spice ginger is the underground rhizome of the ginger plant, known botanically as *Zingiber officinale*. The plant's botanical name is thought to be derived from its Sanskrit name *singabera* which means "horn shaped," a physical characteristic that ginger reflects.

The flesh of the ginger rhizome can be yellow, white or red in colour, depending upon the variety. It is covered with a brownish skin that may either be thick or thin, depending upon whether the plant was harvested when it was mature or young. The ginger rhizome has a firm, yet striated texture and a taste that is aromatic, pungent and hot.

Native to southeastern Asia, a region whose cuisines still feature this wonderfully spicy herb, ginger has been renowned for millennia in many areas throughout the world. Ginger is mentioned in ancient Chinese, Indian and Middle Eastern writings, and has long been prized for its aromatic, culinary and medicinal properties. After the ancient Romans imported ginger from China almost two thousand years ago, its popularity in Europe remained centred in the Mediterranean region until the Middle Ages when its use spread throughout other countries. Although it was a very expensive spice, owing to the fact that it had to be imported from Asia, it was still in great demand. In an attempt to make it more available, Spanish explorers introduced ginger to the West Indies, Mexico and South America, and in the 16th century, these areas began exporting the precious herb back to Europe.

Today, the top commercial producers of ginger include Jamaica, India, Fiji, Indonesia and Australia.

How to Select and Store

Whenever possible, choose fresh ginger over the dried form of the spice since it is not only superior in flavour but contains higher levels of gingerol as well as ginger's active protease (it's anti-inflammatory compound). Fresh ginger root is sold in the produce section of markets. When purchasing fresh ginger root, make sure it is firm, smooth and free of mold. Ginger is generally available in two forms, either young or mature. Mature ginger, the more widely available type, has a tough skin that requires peeling while young ginger, usually only available in Asian markets, does not need to be peeled.

Even through dried herbs and spices like ginger powder are widely available in supermarkets, you may want to explore the local spice stores in your area. Oftentimes, these stores feature an expansive selection of dried herbs and spices that are of superior quality and freshness than those offered in regular markets. Just like with other dried spices, when purchasing dried ginger powder try to select organically grown ginger since this will give you more assurance that it has not been irradiated.

Ginger is also available in several other forms including crystallized, candied and pickled ginger.

Fresh ginger can be stored in the refrigerator for up to three weeks if it is left unpeeled. Stored unpeeled in the freezer, it will keep for up to six months.

Dried ginger powder should be kept in a tightly sealed glass container in a cool, dark and dry place. Alternatively, you can store it in the refrigerator where it will enjoy an extended shelf life of about one year.

Horticulture

Ginger produces clusters of white and pink flower buds that bloom into yellow flowers. Because of its aesthetic appeal and the adaptation of the plant to warm climates, ginger is often used as landscaping around subtropical homes. It is a perennial reed-like plant with annual leafy stems, about a metre (3 to 4 feet) tall. Traditionally, the rhizome is gathered when the stalk withers; it is immediately scalded, or washed and scraped, to kill it and prevent sprouting. The fragrant perisperm of Zingiberaceae is used as sweetmeats by Bantu, also as a condiment and sialogogue.

Health Benefits

Historically, ginger has a long tradition of being very effective in alleviating symptoms of gastrointestinal distress. In herbal medicine, ginger is regarded as an excellent *carminative* (a substance which promotes the elimination of intestinal gas) and *intestinal spasmolytic* (a substance which relaxes and soothes the intestinal tract). Modern scientific research has revealed that ginger possesses numerous therapeutic properties including antioxidant effects, an ability to inhibit the formation of inflammatory compounds, and direct anti-inflammatory effects.

Gastrointestinal Relief

A clue to ginger's success in eliminating gastrointestinal distress is offered by recent double-blind studies, which have demonstrated that ginger is very effective in preventing the symptoms of motion sickness, especially seasickness. In fact, in one study, ginger was shown to be far superior to Dramamine, a commonly used over-the-counter and prescription drug for motion sickness. Ginger reduces all symptoms associated with motion sickness including dizziness, nausea, vomiting, and cold sweating.

Safe and Effective Relief of Nausea and Vomiting During Pregnancy

Ginger's anti-vomiting action has been shown to be very useful in reducing the nausea and vomiting of pregnancy, even the most severe form, *hyperemesis gravidum*, a condition which usually requires hospitalization. In a double-blind trial, ginger root brought about a significant reduction in both the severity of nausea and number of attacks of vomiting in 19 of 27 women in early pregnancy (less than 20 weeks). Unlike antivomiting drugs, which can cause severe birth defects, ginger is extremely safe, and only a small dose is required.

A review of six double-blind, randomized controlled trials with a total of 675 participants, published in the April 2005 issue of the journal, *Obstetrics and Gynecology*,has confirmed that ginger is effective in relieving the severity of nausea and vomiting during pregnancy. The review also confirmed the absence of significant side effects or adverse effects on pregnancy outcomes.

Anti-Inflammatory Effects

Ginger contains very potent anti-inflammatory compounds called *gingerols*. These substances are believed to explain why so many people with osteoarthritis or rheumatoid arthritis experience reductions in their pain levels and improvements in their mobility when they consume ginger regularly. In two clinical studies involving patients who responded to conventional drugs and those who didn't, physicians found that 75% of arthritis patients and 100% of patients with muscular discomfort experienced relief of pain and/or swelling.

Arthritis-related problems with your aging knees? Regularly spicing up your meals with fresh ginger may help, suggests a study published in a recent issue of *Osteoarthritis Cartilage*. In this twelve month study, 29 patients with painful arthritis in the knee (6 men and 23 women ranging in age from 42-85 years) participated in a placebo-controlled, double-blind, crossover study. Patients switched from placebo to ginger or visa versa after 3 months. After six months, the double-blind code was broken and twenty of the patients who wished to continue were followed for an additional six months.

By the end of the first six month period, those given ginger were experiencing significantly less pain on movement and handicap than those given placebo. Pain on movement decreased from a score of 76.14 at baseline to 41.00, while handicap decreased from 73.47 to 46.08. In contrast, those who were switched from ginger to placebo experienced an increase in pain of movement (up to 82.10) and handicap (up to 80.80) from baseline. In the final phase of the study when all patients were getting ginger, pain remained low in those already taking ginger in phase 2, and decreased again in the group that had been on placebo.

Not only did participants' subjective experiences of pain lessen, but swelling in their knees, an objective measurement of lessened inflammation, dropped significantly in those treated with ginger. The mean target knee circumference in those taking ginger dropped from 43.25cm when the study began to 39.36cm by the 12th week. When this group was switched to placebo in the second phase of the study, their knee circumferences increased, while those who had been on placebo but were now switched to ginger experienced a decrease in knee circumference. In the final phase, when both groups were given ginger, mean knee circumference continued to drop, reaching lows of 38.78 and 36.38 in the two groups.

How does ginger work its anti-inflammatory magic? Two other recent studies provide possible reasons.

A study published in the November 2003 issue of *Life Sciences* suggests that at least one reason for ginger's beneficial effects is the free radical protection afforded by one of its active phenolic constituents, 6-gingerol. In this *in vitro* (test tube) study, 6-gingerol was shown to significantly inhibit the production of nitric oxide, a highly reactive nitrogen molecule that quickly forms a very damaging free radical called peroxynitrite. Another study appearing in the November 2003 issue of *Radiation Research* found that in mice, five days treatment with ginger (10 mg per kilogram of body weight) prior to exposure to radiation not only prevented an increase in free radical damage to lipids (fats found in numerous bodily components from cell membranes to cholesterol), but also greatly lessened depletion of the animals' stores of glutathione, one of the body's most important internally produced antioxidants.

A study published in the February 2005 issue of the *Journal of Alternative and Complementary Medicine* sheds further light on the mechanisms of action that underlie ginger's anti-inflammatory effectiveness. In this research, ginger was shown to suppress the pro-inflammatory compounds (cytokines and chemokines) produced by synoviocytes (cells comprising the synovial lining of the joints), chrondrocytes (cells comprising joint cartilage) and leukocytes (immune cells).

Protection against Colourectal Cancer

Gingerols, the main active components in ginger and the ones responsible for its distinctive flavour, may also inhibit the growth of human colourectal cancer cells, suggests research presented at the Frontiers in Cancer Prevention Research, a major meeting of cancer experts that took place in Phoenix, AZ, October 26-30, 2003.

In this study, researchers from the University of Minnesota's Hormel Institute fed mice specially bred to lack an immune system a half milligram of (6)-gingerol three times a week before and after injecting human colourectal cancer cells into their flanks. Control mice received no (6)-gingerol.

Tumors first appeared 15 days after the mice were injected, but only 4 tumors were found in the group of -gingerol-treated mice compared to 13 in the control mice, plus the tumors in the -gingerol group were smaller on average. Even by day 38, one mouse in the (6)-gingerol group still had no measurable tumors. By day 49, all the control mice had been euthanized since their tumors had grown to one cubic centimetre (0.06 cubic inch), while tumors in 12 of the (6)-gingerol treated mice still averaged 0.5 cubic centimetre—half the maximum tumor size allowed before euthanization.

Research associate professor Ann Bode noted, "These results strongly suggest that ginger compounds may be effective chemopreventive and/or chemotherapeutic agents for colourectal carcinomas."

In this first round of experiments, mice were fed ginger before and after tumor cells were injected. In the next round, researchers will feed the mice ginger only after their tumors have grown to a certain size. This will enable them to look at the question of whether a patient could eat ginger to slow the metastasis of a nonoperable tumor. Are they optimistic? The actions of the University of Minnesota strongly suggest they are. The University has already applied for a patent on the use of (6)-gingerol as an anti-cancer agent and has licensed the technology to Pediatric Pharmaceuticals.

Ginger Induces Cell Death in Ovarian Cancer Cells

Lab experiments presented at the 97th Annual Meeting of the American Association for Cancer, by Dr Rebecca Lui and her colleagues from the University of Michigan, showed that gingerols, the active phytonutrients in ginger, kill ovarian cancer cells by inducing apoptosis (programmed cell death) and autophagocytosis (self-digestion).

Ginger extracts have been shown to have both antioxidant, anti-inflammatory and anti-tumor effects on cells. To investigate the latter, Dr Liu examined the effect of a whole ginger extract containing 5% gingerol on a number of different ovarian cancer cell lines.

Exposure to the ginger extract caused cell death in all the ovarian cancer lines studied.

A pro-inflammatory state is thought to be an important contributing factor in the development of ovarian cancer. In the presence of ginger, a number of key indicators of inflammation (vascular endothelial growth factor, interleukin-8 and prostaglandin E2) were also decreased in the ovarian cancer cells.

Conventional chemotherapeutic agents also suppress these inflammatory markers, but may cause cancer cells to become resistant to the action of the drugs. Liu and her colleagues believe that ginger may be of special benefit for ovarian cancer patients because cancer cells exposed to ginger do not become resistant to its cancer-destroying effects. In the case of ovarian cancer, an ounce of prevention—in the delicious form of liberal use of ginger—is an especially good idea. Ovarian cancer is often deadly since symptoms typically do not appear until late in the disease process, so by the time ovarian cancer is diagnosed, it has spread beyond the ovaries. More than 50% of women who develop ovarian cancer are diagnosed in the advanced stages of the disease.

Immune Boosting Action

Ginger can not only be warming on a cold day, but can help promote healthy sweating, which is often helpful during colds and flus. A good sweat may do a lot more than simply assist detoxification. German researchers have recently found that sweat contains a potent germ-fighting agent that may help fight off infections. Investigators have isolated the gene responsible for the compound and the protein it produces, which they have named *dermicidin*. Dermicidin is manufactured in the body's sweat glands, secreted into the sweat, and transported to the skin's surface where it provides protection against invading microorganisms, including bacteria such as *E. coli* and

Staphylococcus aureus (a common cause of skin infections), and fungi, including *Candida albicans.*

Ginger is so concentrated with active substances, you don't have to use very much to receive its beneficial effects. For nausea, ginger tea made by steeping one or two 1/2-inch slices (one 1/2-inch slice equals 2/3 of an ounce) of fresh ginger in a cup of hot water will likely be all you need to settle your stomach. For arthritis, some people have found relief consuming as little as a 1/4-inch slice of fresh ginger cooked in food, although in the studies noted above, patients who consumed more ginger reported quicker and better relief.

Tips for Preparing and Cooking

To remove the skin from fresh mature ginger, peel with a paring knife. The ginger can then be sliced, minced or julienned. The taste that ginger imparts to a dish depends upon when it is added during the cooking process. Added at the beginning, it will lend a subtler flavour while added near the end, it will deliver a more pungent taste.

How to Enjoy}

- Turn up the heat while cooling off by making ginger lemonade. Simply combine freshly grated ginger, lemon juice, cane juice or honey and water.
- Add extra inspiration to your rice side dishes by sprinkling grated ginger, sesame seeds and nori strips on top.
- Combine ginger, soy sauce, olive oil and garlic to make a wonderful salad dressing.
- Add ginger and orange juice to puréed sweet potatoes.
- Add grated ginger to your favourite stuffing for baked apples.
- Spice up your healthy sautéed vegetables by adding freshly minced ginger.

Individual Concerns

Ginger is not a commonly allergenic food and is not known to contain measurable amounts of oxalates or purines.

Nutritional Profile

In addition to the nutrients highlighted in our ratings chart, an in-depth nutritional profile for Ginger is also available. This profile includes information on a full array of nutrients, including carbohydrates, sugar, soluble and insoluble fibre, sodium, vitamins, minerals, fatty acids, amino acids and more.

Introduction to Food Rating System Chart

In order to better help you identify foods that feature a high concentration of nutrients for the calories they contain, we created a Food Rating System. This system allows us to highlight the foods that are especially rich in particular nutrients. The following chart shows the nutrients for which this food is either an excellent, very good, or good source (below the chart you will find a table that explains these qualifications). If a nutrient is not listed in the chart, it does not necessarily mean that the food doesn't contain it. It simply means that the nutrient is not provided in a sufficient amount or concentration to meet our rating criteria. (To view this food's in-depth nutritional profile that includes values for dozens of nutrients - not just the ones rated as excellent, very good, or good - please use the link below the chart.) To read this chart accurately, you'll need to glance up in the top left corner where you will find the name of the food and the serving size we used to calculate the food's nutrient composition. This serving size will tell you how much of the food you need to eat to obtain the amount of nutrients found in the chart. Now, returning to the chart itself, you can look next to the nutrient name in order to find the nutrient amount it offers, the percent Daily Value (DV%) that this amount represents, the nutrient density that we calculated for this food and nutrient, and the rating we established in our rating system. For most of our nutrient ratings, we adopted the government standards for food labelling that are found in the U.S. Food and Drug Administration's "Reference Values for Nutrition Labelling." Read more background information and details of our rating system.

Turmeric

Turmeric, (*Curcuma longa*), perennial herbaceous plant of the ginger family (Zingiberaceae), the tuberous rhizomes, or underground stems, of which have been used from antiquity as a condiment, a textile dye, and medically as an aromatic stimulant. In biblical times it was used as a perfume as well as a spice. In the Middle Ages it was called Indian saffron because of its orange-yellow colour. The rhizome has a pepperlike aroma and a somewhat bitter, warm taste. It is the ingredient that colours and flavours prepared mustard and is used in curry powder, relishes, pickles, spiced butters for vegetables, in fish and egg dishes, and with poultry, rice, and pork. In parts of Asia turmeric water is applied as a cosmetic to lend a golden glow to the complexion.

Native to southern India and Indonesia, turmeric is cultivated on the mainland and in the islands of the Indian Ocean. Production involves a boiling process, which is followed by exposure of the rhizomes to sunlight for five to seven days to dry. Then they are polished by hand rubbing or by rotation in a mounted drum. Dried rhizomes vary from about 2.5 to 7.5 cm (1 to 3 inches) in length. The spice is usually sold in ground form. Distillation yields 1.3 to 5.5 percent essential oil, the main components of which are turmerone and *ar*-turmerone. The colouring matter is curcumin.

Paper tinged with a tincture of turmeric, on addition of alkali, turns from yellow to reddish brown, becoming violet on drying, thus providing a test for alkalinity.

Turmeric (Curcuma langa, L.) is an important condiment and a useful dye, with varied uses in drug and cosmetic industries. It is used medicinally for external application and taken internally as a stimulant. 'Kum-kum', popular with every house wife, is also a by product of turmeric. It finds a place in offerings on religious and ceremonial occasions. A type of starch can also be extracted from a particular type of turmeric. India is one of the leading growers, with about 79,300 ha under this crop, producing 1,43,900 tonnes per annum.

Area of Cultivation

Turmeric is cultivated in Andhra Pradesh, Tamil Nadu, Kerala, Maharashtra, Orissa, West Bengal and the north-eastern states. It is also cultivated in small quantities in Kerala, Bihar, Madhya Pradesh, Rajasthan, Tripura and Assam. In Andhra Pradesh turmeric is grown in Nizamabad. Dugirala and Cuddapah; in Tamil Nadu In Erode and Salem; and in Maharashtra in Sangli, Jat, Kavthemahankal, Karad, Tasgaon and Barsi.

Season

Turmeric can be cultivated in diverse tropical and sub-tropical conditions, to up to 1,500 metres from sea level, with temperatures varying from 20 to 30°C, and a rainfall above 1500 mm. It is a nine month crop sown in July and harvested in April.

Soil

Turmeric thrives in well-drained, fertile, sandy and clayey, black, red or alluvial loams rich in humus and uniform in texture. Rich loamy soils having natural drainage and irrigation facilities are the best. Turmeric cannot stand water stagnation or alkalinity

Rotations

In wet lands turmeric is rotated with paddy, sugarcane, banana, etc. once in 3 or 4 years. In garden lands it is grown in rotation with sugarcane, chilli, onion, garlic, elephant's-foot, yam, vegetables, pulses, wheat, ragi and maize. In Gujarat, it is cultivated as a crop subsidiary to ginger and in other areas with chilli and quick-growing vegetables. Often castor and pigeon-pea are planted on borders and on irrigation channels to provide shade. In West Bengal, turmeric is grown as an inter-crop with manage, jack-tree and litchi and on the west coast with coconut and arecanut.

Cultivation

The land is ploughed 4-6 times to bring the soil to a fine tilth. The land is divided into beds of 1 m width and convenient length with a spacing of 30 cm between the beds for drainage channels. In the case of the irrigated crop, ridges and furrows are prepared and the rhizomes are planed in shallow pits on the top of the ridges. Spacing generally adopted is 45-60 cm between the ridges and 15-22 cm between the plants. In beds, rhizomes are planted 22-35 cm apart in each direction. For sowing both the mother-rhizomes, the fingers are used. The fingers are cut into pieces each 4-5 cm long, and the mother rhizomes are planted as such or split into two; each having at least one sound bud. The seed is sometimes sprouted under moist straw before sowing. Depending on the spacing adopted and the extent of mixed cropping the seed-rate may vary from 2,000-2,500 kg per ha. As an inter-crop in a fruit-garden it may be as low as 400-500 kg per ha. The crop is sown from April to July, depending upon the tract. As in the case of ginger, mulching the bed with green leaves is an important practice beneficial to this crop when planting is done on raised beds. Two or three mulchings are given, the first immediately after planting and the second and the third at intervals of 40-50 days.

Harvesting and Curing

The turmeric crop is ready for harvesting in about 7 to 9 months after sowing depending upon the variety. The aromatica types mature in about 7 months and the longa types in about 9 months and the intermediate types in about 8 months. The dry leaves are cut close to the ground. The land is irrigated, if necessary and ploughed in between the rows if the crop is planted on ridges. Otherwise a crowbar is used. The rhizomes are dug up. The curing quality and the proportion of the cured and dried produce to the green produce depend mainly on the variety. Mother-rhizomes give a higher curing percentage than the fingers. The mother-rhizomes and fingers are separated. If need be, the former is kept for seed and the latter is cured for selling. The green rhizomes are boiled in water till a forth comes out and white fumes appear giving out a characteristic odour. After cooking, the rhizomes would be soft and would yield when pressed between the fingers. The quality of the final product, including the colour and aroma depends largely on the right amount of curing. The boiled rhizomes are spread out on a clean floor and allowed to dry in the sun for about 10-15 days. They are stirred 3 or 4 times to ensure uniform drying. The rounds and fingers are dried separately. The former takes more time to dry. When fully dried the turmeric becomes hard and stiff. The dried turmeric is rubbed against the hard surface of the drying-floor or trampled under feet covered with pieces of gunny cloth. The scales and root bases are separated by win-owing. Clean and big pieces are separated out since they fetch a premium price. The broken bits are taken separately.

Preservation of Seed

Rhizomes for seed are generally heaped in the shade of trees or in well-ventilated sheds and covered with turmeric leaves. Sometimes, the heap is plastered over with earth mixed with cowdung.

Value Added Products

India is the global leader in value-added products development and exports. Germplasm accessions of turmeric were categorized based on levels of essential oil, oleoresins and pungenet principles. Value added products from turmeric include Curcuminoids, dehydrated turmeric powder, oils, and oleoresin.

Onion and Garlic

Onion and garlic are the world famous spice commodities used for flavouring the dishes. Besides culinary purposes, these are considered as valuable medicinal plants. The fungicidal and insecticidal properties of garlic are well established. Dehydrated powder & flakes and paste prepared out of onion and garlic provide rich agro-industrial base for these commodities.

The onion (*Allium cepa*), also known as the bulb onion or common onion, is used as a vegetable and is the most widely cultivated species of the genus *Allium*. This genus also contains several other species variously referred to as onions and cultivated for food, such as the Japanese bunching onion (*A. fistulosum*), the Egyptian onion (*A.* ×*proliferum*), and the Canada onion (*A. canadense*). The name "wild onion" is applied to a number of *Allium* species but *A. cepa* is exclusively known from cultivation and its ancestral wild original form is not known, although escapes from cultivation have become established in some regions. The onion is most frequently a biennial or a perennial plant, but is usually treated as an annual and harvested in its first growing season.

***Figure:** Roots, leaves and developing bulb*

Cultivation

Onions are best cultivated in fertile soils that are well-drained. Sandy loams are good as they are low in sulphur, while clayey soils usually have a high sulphur content and produce pungent bulbs. Onions require a high level of nutrients in the soil. Phosphorus is often present in sufficient quantities but may be applied before planting because of its low level of availability in cold soils.

Nitrogen and potash can be applied at intervals during the growing season, the last application of nitrogen being at least four weeks before harvesting. Bulbing onions are day-length sensitive; their bulbs begin growing only after the number of daylight hours has surpassed some minimal quantity. Most traditional European onions are what is referred to as "long-day" onions, producing bulbs only after 14+ hours of daylight occurs. Southern European and North African varieties are often known as "intermediate day" types, requiring only 12–13 hours of daylight to stimulate bulb formation. Finally, "short-day" onions, which have been developed in more recent times, are planted in mild-winter areas in the fall and form bulbs in the early spring, and require only 11–12 hours of daylight to stimulate bulb formation. Onions are a cool-weather crop and can be grown in USDA zones 3 to 9. Hot temperatures or other stressful conditions cause them to "bolt", meaning that a flower stem begins to grow.

Onions grow well on raised beds or raised rows at least 4 inches high.

Planting

- Select a location with full sun where your onions won't be shaded by other plants.
- Soil needs to be well-drained, loose, and rich in nitrogen; compact soil affects bulb development.
- Till in aged manure or fertilizer the fall before planting. Onions are heavy feeders and need constant nourishment to produce big bulbs.
- At planting time, you can mix in some nitogen fertilizer, too, and side dress every few weeks until the bulbing process begins.
- Seeding? Onion seeds are short-lived. If planting seeds indoors, start with fresh seeds each year. Start seeds indoors about 6 weeks before transplanting.
- Plant onions as soon as the ground can be worked in the spring, usually late March or April. Make sure temperature doesn't go below 20 degrees F.
- For sets or transplants, plant the smaller sets 1 inch deep, with 4 to 5 inches between each plant and in rows 12 to 18 inches apart.
- Think of onions as a leaf crop, not a root crop. When planting onion sets, don't bury them more than one inch under the soil; if more than the bottom third of the bulb is underground, bulb growth can be restricted.
- Practice crop rotation with onions.

Care

- Fertilize every few weeks with nitrogen to get big bulbs. Cease fertilizing when the onions push the soil away and the bulbing process has started. Do not put the soil back around the onions; the bulb needs to emerge above the soil.
- Generally, onions do not need consistent watering if mulch is used. About one inch of water per week (including rain water) is sufficient. If you want sweeter onions, water more.

- Onions will look healthy even if they are bone dry, be sure to water during drought conditions.
- Make sure soil is well-drained. Mulch will help retain moisture and stifle weeds.
- Cut or pull any onions that send up flower stalks; this means that the onions have "bolted" and are done.

Pests and Diseases

Onions suffer from a number of plant disorders. The most serious for the home gardener are likely to be the onion fly, stem and bulb eelworm, white rot and neck rot. Diseases affecting the foliage include rust and smut, downy mildew and white tip disease. The bulbs may be affected by splitting, white rot and neck rot. Shanking is a condition when the central leaves turn yellow and the inner part of the bulb collapses into an unpleasant-smelling slime. Most of these disorders are best treated by removing and burning affected plants. The larvae of the onion leaf miner or leek moth (*Acrolepiopsis assectella*) sometimes attacks the foliage and may burrow down into the bulb.

The onion fly (*Delia antiqua*) lays eggs on the leaves and stems and on the ground close to onion, shallot, leek and garlic plants. The fly is attracted to the crop by the smell of damaged tissue and is liable to occur after thinning. Plants grown from sets are less prone to attack.The larvae tunnel into the bulbs and the foliage wilts and turns yellow. The bulbs are disfigured and rot, especially in wet weather. Control measures may include crop rotation, the use of seed dressings, early sowing or planting and the removal of infested plants.

The onion eelworm (*Ditylenchus dipsaci*), a tiny parasitic soil-living nematode, causes swollen distorted foliage. Young plants are killed and older ones produce soft bulbs. There is no cure and affected plants should be uprooted and burnt. The site should not be used for growing onions again for several years and should also be avoided for growing carrots, parsnips and beans which are also susceptible to the eelworm.

White rot of onions, leeks and garlic is caused by the soil-borne fungus *Sclerotium cepivorum*. As the roots rot, the foliage turns yellow and wilts. The base of the bulbs is attacked and becomes covered by a fluffy white mass of mycelia which later produces small, globular black structures called sclerotia. These resting structures remain in the soil to reinfect a future crop. There is no cure for this fungal disease so affected plants should be removed and destroyed and the ground used for unrelated crops in subsequent years.

Neck rot is a fungal disease affecting onions in storage. It is caused by *Botrytis allii* which attacks the neck and upper parts of the bulb causing a grey mould to develop. The symptoms often first occur where the bulb has been damaged and spread downwards in the affected scales. Large quantities of spores are produced and crust-like sclerotia may also develop. In time a dry rot sets in and the bulb becomes a dry, mummified structure. This disease may be present throughout the growing period but only manifest itself when the bulb is in store. Anti-fungal seed dressings are available and the disease can be minimised by preventing physical damage to the bulbs at harvesting, careful drying and curing of the mature onions and correct storage in a cool, dry place with plenty of circulating air.

Harvest/Storage

- When onions start to mature, the tops become yellow and begin to fall over. At that point, bend the tops down or even stomp on them to speed the final ripening process.
- Loosen the soil to encourage drying, and after a few days turn them up and let them cure on dry ground. Always handle them very carefully—the slightest bruise will encourage rot to set in.
- When tops are brown, pull the onions.
- Be sure to harvest in late summer, before cool weather. Mature onions may spoil in fall weather.
- Allow onions to dry for several weeks before you store them in a root cellar or any other storage area. Spread them out on an open screen off the ground to dry.
- Store at 40 to 50 degrees F (4 to 10 degrees C) in braids or with the stems broken off.
- Mature, dry-skinned bulbs like it cool and dry, so don't store them with apples or potatoes.

Recommended Varieties

Onion varieties are classified into two categories: Long-day, best in the North, and short-day, best for the South.

Garlic

Allium sativum, commonly known as garlic, is a species in the onion genus, *Allium*.

Its close relatives include the onion, shallot, leek, chive, and rakkyo. With a history of human use of over 7,000 years, garlic is native to central Asia, and has long been a staple in the Mediterranean region, as well as a frequent seasoning in Asia, Africa, and Europe. It was known to Ancient Egyptians, and has been used for both culinary and medicinal purposes.

For a small vegetable, garlic (*Allium sativum*) sure has a big, and well deserved, reputation. And although garlic may not always bring good luck, protect against evil, or ward off vampires, characteristics to which it has been assigned folklorically, it is guaranteed to transform any meal into a bold, aromatic, and healthy culinary experience. Garlic is a member of the Lily family and is a cousin to onions, leeks and chives.

Garlic is arranged in a head, called a "bulb," which averages about 2 inches in height and diametre and consists of numerous small separate cloves. Both the cloves and the entire bulb are encased in paper-like sheathes that can be white, off-white, or have a pink/purple hue. Although garlic cloves have a firm texture, they can be easily cut or crushed. The taste of garlic is like no other—it hits the palate with a hot pungency that is shadowed by a very subtle background sweetness. While elephant garlic has larger cloves, it is more closely related to the leek and therefore does not offer the full health benefits of regular garlic.

Fresh, dried and powdered garlic are available in markets throughout the year, however, fresh varieties from California are in season from June through December.

Cultivation

Garlic is easy to grow and can be grown year-round in mild climates. While sexual propagation of garlic is indeed possible, nearly all of the garlic in cultivation is propagated asexually, by planting individual cloves in the ground. In cold climates, cloves are planted in the autumn, about six weeks before the soil freezes, and harvested in late spring. The cloves must be planted at sufficient depth to prevent freeze/thaw which causes mold or white rot Garlic plants are usually very hardy, and are not attacked by many pests or diseases. Garlic plants are said to repel rabbits and moles. Two of the major pathogens that attack garlic are nematodes and white rot disease, which remain in the soil indefinitely after the ground has become infected. Garlic also can suffer from pink root, a typically nonfatal disease that stunts the roots and turns them pink or red.

Garlic plants can be grown closely together, leaving enough space for the bulbs to mature, and are easily grown in containers of sufficient depth. Garlic does well in loose, dry, well drained soils in sunny locations, and is hardy throughout USDA climate zones 4–9. When selecting garlic for planting, it is important to pick large bulbs from which to separate cloves. Large cloves, along with proper spacing in the planting bed, will also improve bulb size. Garlic plants prefer to grow in a soil with a high organic material content, but are capable of growing in a wide range of soil conditions and pH levels.

There are different types or subspecies of garlic, most notably hardneck garlic and softneck garlic. The latitude where the garlic is grown affects the choice of type as garlic can be day-length sensitive. Hardneck garlic is generally grown in cooler climates; softneck garlic is generally grown closer to the equator.

Garlic scapes are removed to focus all the garlic's energy into bulb growth. The scapes can be eaten raw or cooked.

Production Trends

Garlic is grown globally, but China is by far the largest producer of garlic, with approximately 10.5 million tonnes (23 billion pounds) grown annually, accounting for over 77% of world output. India (4.1%) and South Korea (2%) follow, with Egypt and Russia (1.6%) tied in fourth place and the United States (where garlic is grown in every state except for Alaska) in sixth place (1.4%). This leaves 16% of global garlic production in countries that each produce less than 2% of global output. Much of the garlic production in the United States is centred in Gilroy, California, which calls itself the "garlic capital of the world".

Properties

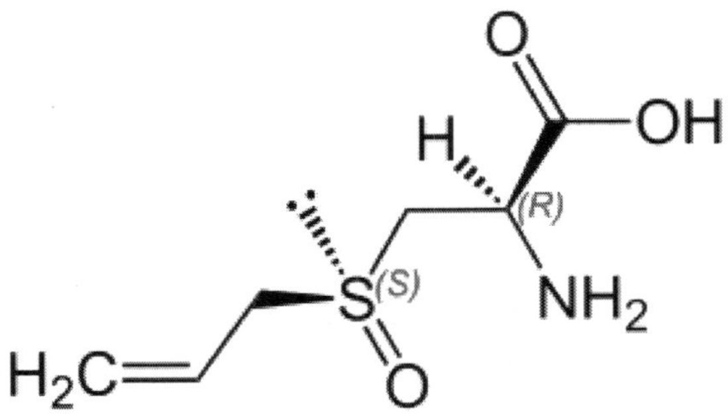

***Figure:** Alliin, a sulfur-containing compound found in garlic.*

When crushed, *Allium sativum* yields allicin, an antibiotic and antifungal compound (phytoncide) discovered by Chester J. Cavallito and colleagues in 1944. Fresh or crushed garlic also affords the sulfur-containing compounds alliin, ajoene, diallyl polysulfides, vinyldithiins, *S*-allylcysteine, and enzymes, B vitamins, proteins, minerals, saponins, flavonoids, and Maillard reaction products, which are not sulfur-containing compounds. Furthermore, a phytoalexin (allixin) was found, a nonsulfur compound with a ã-pyrone skeleton structure with antioxidant effects, antimicrobial effects, antitumor promoting effects, inhibition of aflatoxin B2 DNA binding, and neurotrophic effects. Allixin showed an antitumor promoting effect *in vivo*, inhibiting skin tumor formation by TPA and DMBA initiated mice. Analogs of this compound have exhibited antitumor promoting effects in *in vitro* experimental conditions. So allixin and/or its analogs may be useful compounds for cancer prevention.

The composition of the bulbs is approximately 84.09% water, 13.38% organic matter, and 1.53% inorganic matter, while the leaves are 87.14% water, 11.27% organic matter, and 1.59% inorganic matter.

The phytochemicals responsible for the sharp flavour of garlic are produced when the plant's cells are damaged. When a cell is broken by chopping, chewing, or crushing, enzymes stored in cell vacuoles trigger the breakdown of several sulfur-containing compounds stored in the cell fluids (cytosol). The resultant compounds are responsible for the sharp or hot taste and strong smell of garlic. Some of the compounds are unstable and continue to react over time. Among the members of the onion family, garlic has by far the highest concentrations of initial reaction products, making garlic much more potent than onion, shallot, or leeks. Although many humans enjoy the taste of garlic, these compounds are believed to have evolved as a defensive mechanism, deterring animals such as birds, insects, and worms from eating the plant.

A large number of sulfur compounds contribute to the smell and taste of garlic. Allicin has been found to be the compound most responsible for the "hot" sensation of raw garlic. This chemical opens thermo-transient receptor potential channels that are responsible for the burning sense of heat in foods.

The process of cooking garlic removes allicin, thus mellowing its spiciness. Allicin, along with its decomposition products diallyl disulfide and diallyl trisulfide, are major contributors to the characteristic odor of garlic, while other allicin-derived compounds, such as vinyldithiins and ajoene show beneficial in vitro biological activity. Because of its strong odor, garlic is sometimes called the "stinking rose". When eaten in quantity, garlic may be strongly evident in the diner's sweat and garlic breath the following day. This is because garlic's strong-smelling sulfur compounds are metabolized, forming allyl methyl sulfide. Allyl methyl sulfide (AMS) cannot be digested and is passed into the blood. It is carried to the lungs and the skin, where it is excreted. Since digestion takes several hours, and release of AMS several hours more, the effect of eating garlic may be present for a long time.

The well-known phenomenon of "garlic breath" is allegedly alleviated by eating fresh parsley. The herb is, therefore, included in many garlic recipes, such as *pistou*, *persillade*, and the garlic butter spread used in garlic bread. However, since the odor results mainly from digestive processes placing compounds such as AMS in the blood, and AMS is then released through the lungs over the course of many hours, eating parsley provides only a temporary masking. One way of accelerating the release of AMS from the body is the use of a sauna.

Because of the AMS in the bloodstream, it is believed by some to act as a mosquito repellent, but no clinically reported evidence suggests it is actually effective.

Abundant sulfur compounds in garlic are also responsible for turning garlic green or blue during pickling and cooking. Under these conditions (*i.e.* acidity, heat) the sulfur-containing compound alliinase react with common amino acids to make pyrroles, clusters of carbon-nitrogen rings. These rings can be linked together into polypyrrole molecules. Ring structures absorb particular wavelengths of light and thus appear coloured. The two-pyrrole molecule looks red, the three-pyrrole molecule looks blue and the four-pyrrole molecule looks green (like chlorophyll, a tetrapyrrole). Like chlorophyll, the pyrrole pigments are safe to eat.

Dahlia

Dahlia is a genus of bushy, tuberous, herbaceous perennial plants native mainly in Mexico, but also Central America, and Colombia. A member of the Asteraceae or Compositae, dicotyledonous plants, related species include the sunflower, daisy, chrysanthemum and zinnia. There are at least 36 species of dahlia, with hybrids commonly grown as garden plants. Flower forms are variable, with one head per stem; these can be as small as 2 in (5.1 cm) diametre or up to 1 ft (30 cm) ("dinner plate"). This great variety results from dahlias being octoploids—that is, they have eight sets of homologous chromosomes, whereas most plants have only two. In addition, dahlias also contain many transposons—genetic pieces that move from place to place upon an allele—which contributes to their manifesting such great diversity.

The stems are leafy, ranging in height from as low as 12 in (30 cm) to more than 6–8 ft (1.8–2.4 m). The majority of species do not produce scented flowers or cultivars. Like most plants that do not attract pollinating insects through scent, they are brightly coloured, displaying most hues, with the exception of blue.

The dahlia was declared the national flower of Mexico in 1963. The tubers were grown as a food crop by the Aztecs, but this use largely died out after the Spanish Conquest. Attempts to introduce the tubers as a food crop in Europe were unsuccessful.

Classification

Since 1789 when Cavanilles first flowered the dahlia in Europe, there has been an ongoing effort by many growers, botanists and taxonomists, to determine the development of the dahlia to modern times. At least 85 species have been reported: approximately 25 of these were first reported from the wild, the remainder appeared in gardens in Europe. They were considered hybrids, the results of crossing between previously reported species, or developed from the seeds sent by Humboldt from Mexico in 1805, or perhaps from some other undocumented seeds that had found their way to Europe. Several of these were soon discovered to be identical with earlier reported species, but the greatest number are new varieties. Morphological variation is highly pronounced in the dahlia. William John Cooper Lawrence, who hybridized hundreds of families of dahlias in the 1920s, stated: "I have not yet seen any two plants in the families I have raised which were not to be distinguished one from the other. Constant reclassification of the 85 reported species has resulted in a considerably smaller number of distinct species, as there is a great deal of disagreement today between systematists over classification.

In 1829, all species growing in Europe were reclassified under an all-encompassing name of *D. variabilis, Desf.* Through the interspecies cross of the Humboldt seeds and the Cavanilles species, 22 new species were reported by that year, all of which had been classified in different ways by several different taxonomists, creating considerable confusion as to which species was which.

In 1830 William Smith suggested that all dahlia species could be divided into two groups for colour, red-tinged and purple-tinged(92). In investigating this idea Lawrence(58) determined that with the exception of *D. variabilis*, all dahlia species may be assigned to one of two groups for flower-colour: Group I (ivory-magenta) or Group II (yellow-orange-scarlet).

Cultivation

In 1846 the Caledonia Horticultural Society of Edinburgh, offered a prize of 2,000 pounds to the first person succeeding in producing a blue dahlia. This has to date not been accomplished. While dahlias produce anthocyanin, an element necessary for the production of the blue, to achieve a true blue colour in a plant, the anthocyanin delphinidin needs six hydroxyl groups. To date dahlias have only developed five, so the closest that breeders have come to achieving a "blue" specimen are variations of mauve, purples and lilac hues.

Dahlias grow naturally in climates which do not experience frost (the tubers are hardy to USDA Zone 8), consequently they are not adapted to withstand sub-zero temperatures. However their tuberous nature enables them to survive periods of dormancy, and this characteristic means that gardeners in temperate climates with frosts can grow dahlias successfully, provided the tubers are lifted from the ground and stored in cool yet frost-free conditions during the winter.

Planting the tubers quite deep (10 – 15 cm) also provides some protection. When in active growth, modern dahlia hybrids perform most successfully in well-watered yet free-draining soils, in situations receiving plenty of sunlight. Taller cultivars usually require some form of staking as they grow, and all garden dahlias need deadheading regularly, once flowering commences.

Pests and Diseases

Slugs and snails are serious pests in some parts of the world, particularly in spring when new growth is emerging through the soil. Earwigs can also disfigure the blooms. The other main pests likely to be encountered are aphids (usually on young stems and immature flower buds), red spider mite (causing foliage mottling and discolouration, worse in hot and dry conditions) and capsid bugs (resulting in contortion and holes at growing tips). Diseases affecting dahlias include powdery mildew, grey mould (*Botrytis cinerea*), verticillium wilt, dahlia smut (*Entyloma calendulae* f. *dahliae*), phytophthora and some plant viruses. Dahlias are a source of food for the larvae of some Lepidoptera species including Angle Shades, Common Swift, Ghost Moth and Large Yellow Underwing.

Food and Medicinal Uses

Today the dahlia is still considered one of the native ingredients in Oaxacan cuisine; several cultivars are still grown especially for their large, sweet potato-like tubers. Dacopa, an intense mocha-tasting extract from the roasted tubers, is used to flavour beverages throughout Central America.

In Europe and America, prior to the discovery of insulin in 1923, diabetics—as well as consumptives—were often given a substance called *Atlantic starch* or *diabetic sugar*, derived from inulin, a naturally occurring form of fruit sugar, extracted from dahlia tubers. Inulin is still used in clinical tests for kidney functionality.

Coleus

Coleus was a genus of flowering plants in the family Lamiaceae. In recent classifications, the genus is no longer recognized, and the formerly included species are instead placed in the genera *Plectranthus* and *Solenostemon*. Because the type species, *Coleus amboinicus* is now placed in *Plectranthus*, *Coleus* is regarded as a synonym of *Plectranthus*. The term "coleus" is often used as a common name for species formerly place in the genus *Coleus* that are cultivated as ornamental plants, particularly *Coleus blumei* (*Solenostemon scutellarioides*), which is popular as a garden plant for its brightly coloured foliage.

Figure: *Hybrid coleus*

Tapioca

Tapioca (Portuguese pronunciation: This species is native to Northern Brazil but spread throughout the South American continent. The plant was carried by Portuguese and Spanish explorers to most of the West Indies, and continents of Africa and Asia, including the Philippines and Taiwan; it is cultivated worldwide.

In Brazil, the plant (cassava) is named *mandioca*, while its starch is called *tapioca*. The name *tapioca* is derived from the word *tipi'óka*, its name in the Tupí language, which was spoken by the natives when the Portuguese first arrived in the Northeast of Brazil. This Tupí word refers to the process by which the cassava starch is made edible. As the food and word were taken to other areas, "tapioca" was also applied to similar preparations made with other esculents.

In the Philippines, tapioca is often confused with *sago*, as the sap of the sago palm is often part of its preparation. A coarsely granular substance obtained by heating, and thus partly changing, the moistened starch obtained from the roots of the cassava. It is much used in puddings and as a thickening for soups. In India, the term "tapioca-root" is used to represent the root of the plant (cassava), rather than the starch. It is widely named as (*kappa*) in Malayalam. In Vietnam, it is called *bỪt nãng*. In the past, the starch was extracted from Maranta arundinacea (*Mì tinh, hoàng tinh*). In Indonesia, it is called *singkong*. In Malaysia, it is called *Ubi Kayu*.

Tapioca is a staple food in some regions, and it is used worldwide as a thickening agent in various foods. It is a gluten-free food.

Tapioca is one of the purest forms of starch food, and the production varies from region to region.

The cassava plant has either red or green branches with blue spindles on them. The root of the green-branched variant requires treatment to remove linamarin, a cyanogenic glycoside occurring naturally in the plant, which otherwise may be converted into cyanide. Konzo (also called mantakassa) is a paralytic disease associated with several weeks of almost exclusive consumption of insufficiently processed bitter cassava.

In the North and Northeast of Brazil, traditional community based production of tapioca is a by-product of manioc flour production from cassava roots. In this process, the manioc (after treatment to remove toxicity) is ground to a pulp with a small hand- or diesel-powered mill. This masa is then squeezed to dry it out. The wet masa is placed in a long woven tube called a *tipiti*. The top of the tube is secured while a large branch or lever is inserted into a loop at the bottom and used to stretch the entire implement vertically, squeezing a starch-rich liquid out through the weave and ends. This liquid is collected and the water allowed to evaporate, leaving behind a fine-grained tapioca powder similar in appearance to corn starch.

Commercially, the starch is processed into several forms: hot soluble powder, meal, pre-cooked fine/coarse flakes, rectangular sticks, and spherical "pearls". Pearls are the most widely available shape; sizes range from about 1 mm to 8 mm in diametre, with 2–3 mm being the most common.

Flakes, sticks, and pearls must be soaked well before cooking, in order to rehydrate, absorbing water up to twice their volume. After rehydration, tapioca products become leathery and swollen. Processed tapioca is usually white, but sticks and pearls may be coloured. Since old times, the most common colour applied to tapioca has been brown, but recently pastel colours have been available. Tapioca pearls are generally opaque when raw, but become translucent when cooked in boiling water.

Brazil in South America, Thailand in Asia, and Nigeria in Africa are the world's largest producers of cassava. Currently, Thailand accounts for about 60% of worldwide exports.

Uses

South America

Brazil: In Brazilian cuisine, tapioca is used for different types of meals. In beiju (or biju), the tapioca is moistened, strained through a sieve to become a coarse flour, then sprinkled onto a hot griddle or pan, where the heat makes the starchy grains fuse into a flatbread which resembles a grainy pancake. Then it may be buttered and eaten as a toast (its most common use as a breakfast dish), or it may be filled or topped with either *salgados* (salty ingredients) or *doces* (sweet ingredients), which define the kind of meal the tapioca is used for: breakfast/dinner, or dessert. Choices for fillings range from butter, cheese, ham, bacon, various kinds of meat, chocolate, fruits such as ground coconut, condensed milk, chocolate with sliced pieces of banana or strawberry, among others. This kind of tapioca dish is usually served warm.

A regional dessert called *sagu* is also made in Southern Brazil from tapioca pearls cooked with cinnamon and cloves in red wine. The cassava root is known by different names throughout the country: *mandioca* in the North, Central-West and in São Paulo; tapioca or *macaxeira* in the Northeast; *aipim* in the Southeast (especially in Rio de Janeiro).

The fine-grained tapioca starch is called *polvilho,* and it is classified as either "sweet" or "sour". Sour *polvilho* is commonly used in dishes such as *pão de queijo* or "cheese bread", in which the starch is mixed with a certain type of cheese similar to parmigiano (parmesan), eggs and butter and baked in the oven. The final result is an aromatic, chewy and elastic kind of bread that is ubiquitous across the country. Sour cassava flour is mixed into mashed beans to make the dish *tutu de feijão*.

Other Locations

In Colombia and Venezuela, arepas may be made with tapioca flour rather than cornmeal. Tapioca arepas probably predate cornmeal arepas; among traditional cultures of the Caribbean the name for them is casabe. Throughout both Spanish and Portuguese South America, the tapioca, or yuca, starch is used to make regional variations of the baked cheese bun, known locally as *pandebono, pan de yuca, pão de queijo, chipá*, or *cuñapé*, among other names.

The whole unprocessed cassava root also has a number of culinary uses throughout South America.

North America

While frequently associated with tapioca pudding, a dessert in the United States, tapioca is also used in other courses. Bubble tea, made with tapioca pearls, is gaining popularity in cities with large Asian populations. People on gluten-free diets can eat bread made with tapioca flour (although these individuals have to be careful, as some tapioca flour has wheat added to it). Tapioca is also used as an ingredient in the Canadian Daiya brand cheese substitute.

West Indies

Tapioca was used by the first inhabitants of the West Indies as a staple food from which they made main dishes, such as pepper pot, and also used it to make alcohol. They used it for teeth cleaning, and it is still used as a base locally for toothpaste. In the 21st century, it is still a very popular food in the islands, used as a provision cooked with meats or fish, and in desserts such as cassava pone.

Asia

In various Asian countries, including Indonesia, China, Thailand, Vietnam, India, Pakistan, Bangladesh, Myanmar, Philippines, Malaysia, and Taiwan, tapioca pearls are widely used and are known as sagudana, sabudana or shabudana (pearl sago) or “sabba akki”. It has religious importance in certain communities and used as a staple food for fasting. The pearls (sagudana or shabudana/sabudana) are used to make snacks. Tapioca pearls are essential ingredients for Taiwanese Bubble Tea.

Other Root Tubers

A tuberous root or storage root, is a modified lateral root, enlarged to function as a storage organ. The enlarged area of the root-tuber, or storage root, can be produced at the end or middle of a root or involve the entire root. It is thus different in origin but similar in function and appearance to a stem tuber. Examples of plants with notable tuberous roots include the sweet potato, cassava, yam and dahlia.

Root tubers are perennating organs, thickened roots that store nutrients over periods when the plant cannot actively grow, thus permitting survival from one year to the next. The massive enlargement of secondary roots typically represented by Sweet Potato (*Ipomoea batatas*), have the internal and external cell and tissue structures of a normal root, they produce adventitious roots and stems which again produce adventitious roots.

In root-tubers, there are no nodes and internodes or reduced leaves. Root tubers have one end called the proximal end, which is the end that was attached to the old plant; this end has crown tissue that produces buds which grow into new stems and foliage. The other end of the root tuber is called the distal end, and it normally produces unmodified roots. In stem tubers the order is reversed, with the distal end producing stems. Tuberous roots are biennial in duration: the first year the plant produces root-tubers, and at the end of the growing season, the plant shoots often die, leaving the newly generated tubers. The next growing season, the root-tubers produce new shoots. As the shoots of the new plant grow, the stored reserves of the root-tuber are consumed in the production of new roots, stems, and reproductive organs; any remaining root tissue dies concurrently to the plant's regeneration of next generation of root-tubers.

Hemerocallis fulva plus a number of Daylily hybrids have large root-tubers, *H. fulva* spreads by underground stolons that end with a new fan that grows roots that produce thick root tubers and then send out more stolons.

Root tubers, along with other storage tissues that plants produce, are consumed by animals as a rich source of nutrients. The root-tubers of Arrowhead plants of the genus *Sagittaria* are eaten by ducks.

Plants with root tubers are propagated in late summer to late winter by digging up the tubers and separating them, making sure that each piece has some crown tissue for replanting.

Pest Management

Integrated Pest Management (IPM) is an ecosystem approach to crop production and protection that combines different management strategies and practices to grow healthy crops and minimize the use of pesticides.

Pest management can be one of the greatest challenges to the home gardener. Yard pests include weeds, insects, diseases, and some species of wildlife. Weeds are plants that are growing out of place. Insect pests include an enormous number of species from tiny thrips, that are nearly invisible to the naked eye, to the large larvae of the tomato hornworm.

Diseases are caused by fungi, bacteria, viruses, and other organisms, some of which are only now being classified. Poor plant nutrition and misuse of pesticides also can cause injury to plants. Slugs, mites, and many species of wildlife such as rabbits, deer, and crows can be extremely destructive.

Identify the Problem

Careful identification of the problem is essential before control practices can be used. Some insect damage may appear to be a disease, especially if no visible insects are present. Nutrient problems may also mimic diseases. Herbicide damage resulting from misapplication of chemicals also can be mistaken for other problems.

What to Look For

Insects and Mites: All insects have six legs, but other than that they are extremely variable. They include such organisms as beetles, flies, bees, ants, moths, and butterflies. Mites and spiders have eight legs—they are not insects. But for the purposes of this tip sheet, they will be considered as insects.

Insects damage plants in several ways. The most visible damage is chewed plant leaves and flowers. Many pests are visible and can be readily identified, including the Japanese beetle, Colorado potato beetle, and numerous species of caterpillars such as tent caterpillars and tomato hornworms. Other chewing insects, however, such as cutworms (which are caterpillars) come out at night to eat, and burrow into the soil during the day. These are much harder to identify but should be considered if young plants seem to disappear overnight or are found cut off at ground level.

Sucking insects are extremely common and can be very damaging. These insects insert their mouth parts into the plant tissues and suck out the plant juices. They also may carry diseases that they spread from plant to plant as they move about the yard. You may suspect that these insects are present if you notice misshapen plant leaves or flower petals. Often the younger leaves will appear curled or puckered. Flowers developing from the buds may only partially develop. Look on the underside of the leaves as that is where many species tend to gather. Common sucking insects include leafhoppers, aphids, mealy bugs, thrips, and mites.

Other insects cause damage by boring into stems, fruits, and leaves. They may disrupt the plant's ability to transport water. They also create opportunities for disease organisms to attack the plants. You may suspect the presence of boring insects if you see small accumulations of sawdust-like material on plant stems or fruits. Common examples of boring insects include squash vine borers and corn borers.

Diseases

Plant disease identification is extremely difficult. In some cases, only laboratory analysis can conclusively identify some diseases. Disease organisms injure plants in several ways. Some attack leaf surfaces and limit the plant's ability to carry on photosynthesis. Other organisms produce substances that clog plant tissues that transport water and nutrients. Other disease organisms produce toxins that kill the plant or replace plant tissue with their own.

Symptoms associated with plant diseases may include the presence of mushroom-like growths on trunks of trees; leaves with a grayish mildewy appearance; spots on leaves, flowers, and fruits; sudden wilting or death of a plant or branch; sap exuding from branches or trunks of trees; and stunted growth.

Misapplication of pesticides and nutrients, air pollutants, and other environmental conditions such as flooding and freezing can also mimic some disease problems. Yellowing or reddening of leaves and stunted growth may indicate a nutritional problem. At first glance, blossom end rot of tomato, in which the bottom of the tomato turns black, might appear to be a disease caused by some pathogen. It is actually caused by the plant's inability to take up calcium quickly enough during periods of rapid growth. Prevent this problem with adequate moisture—adding more calcium is of no benefit! Leaf curling or misshapen growth may be a result of herbicide application.

Pest Management Practices

Preventing pests should be your first goal, although it is unlikely that you will be able to avoid all pest problems because some plant seeds and disease organisms may lay dormant in the soil for years.

Diseases need three elements to become established: the disease organism, a susceptible species, and the proper environmental conditions. Some disease organisms can live in the soil for years; other organisms are carried in infected plant material that falls to the ground. Some disease organisms are carried by insects. Good sanitation will help limit some problems. Planting resistant varieties of plants prevents many diseases. Rotating annual crops in a garden also prevents some diseases.

You will likely have the most opportunity to alter the environment in favour of the plant and not the disease. Healthy, vigorous lawn and garden plants have a higher resistance to pests. Plants that have adequate, but not excessive, nutrients are better able to resist attacks from both diseases and insects. Excessive rates of nitrogen often result in extremely succulent vegetative growth and can make plants more susceptible to insect and disease problems, as well as decrease their winter hardiness. Proper watering and spacing of plants limits the spread of some diseases. Some disease species require free standing water in which to spread, while other species just need high humidity. Proper spacing provides good aeration around plants. Trickle irrigation where water is applied to the soil and not the plant leaves may be helpful.

Barriers may be effective to exclude some pests. Mulching is effective against weeds. Fences can limit damage from rabbits. Row covers may prevent insect damage on young vegetable plants. Netting can be applied to small fruit trees and berries to limit damage from birds.

Integrated Pest Management (IPM)

It is difficult, if not impossible, to prevent all pest problems every year. If your best prevention efforts have not been entirely successful, you may need to use some control methods. Integrated Pest Management (IPM) relies on several techniques to keep pests at acceptable population levels without excessive use of chemical controls. The basic principles of IPM include monitoring (scouting), determining tolerable injury levels (thresholds), and applying appropriate strategies and tactics. Unlike other methods of pest control where pesticides are applied on a rigid schedule, IPM applies only those controls that are needed, when they are needed, to control pests that will cause more than a tolerable level of damage to the plant.

Monitoring is essential for a successful IPM programme. Check your plants regularly. Look for signs of damage from insects and diseases as well as indications of adequate fertility and moisture. Early identification of potential problems is essential.

There are thousands of insects in the garden, many of which are harmless or even beneficial. Proper identification is needed before control strategies can be adopted. It is important to recognize the different stages of insect development for several reasons. The caterpillar eating your plants may be the larvae of the butterfly you were trying to attract. The small larvae with six spots on its back is probably the young of the ladybug, a very beneficial insect. Some control practices are most effective on young insects. Different stages may also be more damaging than others.

It is not necessary to kill every insect, weed, or disease organism to have a healthy yard. This is where the concept of thresholds comes in. The economic threshold is the point where the damage caused by the pest exceeds the cost of control. In a home garden, this can be difficult to determine. What you are growing and how you intend to use it will determine how much damage you are willing to tolerate. Remember that larger plants, especially those close to harvest, can tolerate more damage than a tiny seedling. A few flea beetles on a radish seedling may warrant control whereas numerous Japanese beetles eating the leaves of beans close to harvest may not.

If the threshold level for control has been exceeded, you may need to employ control strategies. Strategies can be discussed with the Cooperative Extension Service, garden centres, or nurseries.

Control Strategies

Mechanical/physical Controls

Insects: Many insects can be removed by hand. This method is preferable if a few, large insects are causing the problem. Simply remove the insect from the plant and drop it into a container of soapy water or vegetable oil. Caution: some insects have spines or excrete oily substances that can cause injury to humans. Use caution when handling unfamiliar insects. Wear gloves or remove insects with tweezers.

Many insects can be removed from plants by spraying water from a hose or sprayer. Small vacuums can be used to suck up insects. Traps can be used effectively for some insects. These come in a variety of styles depending on the insect to be caught. Many traps rely on the use of pheromones—naturally occurring chemicals produced by the insects and used to attract the opposite sex during mating. They are extremely specific for each species and, therefore, will not harm beneficial species. One caution with traps is that they may actually draw more insects into your yard. You should not place them directly in the garden. Other traps are more generic and will attract numerous species. These include such things as yellow and blue sticky cards. Different insects are attracted to different colours. Sticky cards also can be used effectively to monitor insect pests.

Weeds

Hoeing, pulling, and mulching are the most effective physical control methods for weeds. Weeding is most important while plants are small. Well established plants often can tolerate competition from weeds.

Diseases

Removal of diseased material limits the spread of some diseases. Clean up litter dropped from diseased plants. Prune diseased branches on trees and shrubs. When pruning diseased trees and shrubs, disinfect your pruners between cuts with a solution of chlorine bleach to avoid spreading the disease from plant to plant. Control insects known to spread plant diseases.

Other Pests

Fences, netting, and tree trunk guards can be extremely successful in limiting damage from small mammals and birds. Numerous traps are also available to catch or kill some animals. Caution: In many states it is illegal to move wildlife, including squirrels. Traps may also catch animals other than the ones targeted. Check local regulations before trapping.

Diatomaceous earth, a powder-like dust made of tiny marine organisms called diatoms, can be used to reduce damage from soft-bodied insects and slugs. Spread this material on the soil—it is sharp and cuts or irritates these soft organisms. It is harmless to other organisms. Shallow dishes of beer can be used to trap slugs.

Biological Controls

Biological controls are nature's way of regulating populations. Biological controls rely on predators and parasites to keep organisms under control. Many of our present pest problems result from the loss of predator species.

Beneficial Insect	***Controls***
Green lacewings	aphids, mealy bugs, thrips, spider mites
Ladybugs	aphids, Colorado potato beetle
Praying mantis	almost any insect
Ground beetles	caterpillars that attack trees and shrubs
Parasitic nematodes(tiny worm-like organisms)	grubs, beetles, cutworms, army worms
Trichogramma wasp(extremely small, non-stinging wasps)	corn borer, cabbage looper, other worms
Seedhead weevils and other beetles	weeds

Other biological controls include birds and bats that eat insects. A single bat can eat up to 600 mosquitoes an hour. Many bird species eat insect pests on trees and in the garden. Bacillus thuringiensis (Bt) is a bacteria that specifically attacks larvae of some insect pests including white grubs in the lawn and Japanese beetles. This bacteria is harmless to desirable species.

Chemical Controls

When using biological controls, be very careful with pesticides. Most common pesticides are broad spectrum in that they kill a wide variety of organisms. Spray applications of insecticides are likely to kill numerous beneficial insects as well as the pests. Herbicides applied to weed species may drift in the wind or vaporize in the heat of the day and injure non-targeted plants. Runoff of pesticides can pollute water. Many pesticides are toxic to humans as well as pets and small animals that may enter your yard.

Some common, non-toxic household substances are as effective as many more toxic compounds. A few drops of dishwashing detergent mixed with water and sprayed on plants is extremely effective in controlling many soft-bodied insects such as aphids and whiteflies. Crushed garlic mixed with water may control certain insects. A baking soda solution has been shown to help control some fungal diseases on roses.

When using pesticides, follow label directions carefully. Altering the rate of application or increasing the frequency of application can injure desirable plant and animal species. Spot applications of the pesticide to the targeted pest can reduce the amount used and help reduce the risk of injury to non-targeted species. Do not apply on windy days. Read the label for information on other environmental conditions such as temperature and rain that may influence the pesticide's effectiveness. Be aware that many so-called "organic" pesticides may be just as toxic as the synthetic chemical products.

On the Farm

Proper pest management on the farm involves a variety of practices, such as rotating crops to reduce disease and insect problems, and establishing tall grass hedges to provide habitat for beneficial insects. Many farmers now monitor their fields regularly — a practice called scouting — to keep track of insect and weed populations. Only when populations reach a level where an unacceptable amount of damage is likely are direct control measures initiated. When pesticides are necessary, farmers fill and clean tanks away from water sources, mix only necessary amounts, and apply only to land where problems exist.

Many farmers are using IPM techniques for effective and economic control of pests. IPM programmes have been developed for numerous crops including apples, grapes, corn, and alfalfa. Based on monitoring of specific pests, farmers apply only those pesticides needed for adequate control.

Sources of Critical Inputs for the Management of Pest and Diseases

Many examples have shown a reduced reliance on pesticides, effectively diminishing their use. Lobbying and advocacy actions have led to new rules and regulations in many countries, forbidding the commercialisation and use of the most toxic products, and at the same time greatly increasing the awareness of farmers and consumers.

But the problems persist, and these are not limited to large scale farmers or to intensive agriculture situations. IPM has had many successes, but the magnitude of the problem which farmers currently face forces us to look once again at pests and diseases and at their management.

While it may be necessary to critically assess the extension methods followed during these last two decades, it seems equally relevant to stress and emphasise again one of the basic ideas behind the "different pest management system . Pests and diseases are not an isolated part of agriculture, but rather a symptom of a broader problem, and need thus to be seen –and managed– accordingly.

A major component of managing pests and diseases was the relationship between pest incidence, soil fertility and soil organic matter content: "because of bad soil condition and consequent bad growth, plants become more susceptible to pest attacks".

Originally designed as a technical approach to reduce the number of pesticide applications, IPM developed into a comprehensive methodology, based on farmers' better understanding of their own agroecosystems. IPM first considered the integration of control methods and target pests as a combination of biological control methods, host plant resistance, cultural control and selective chemical control. When it was realised that many agricultural practices influence pest incidence, farm management was also added to the equation as an important part of an IPM approach, together with management of natural resources.

Gradually, social aspects were also included, paying attention to women's roles in pest and disease management, to the role of local organisations and to the importance of indigenous knowledge in the process of developing the necessary skills and confidence to make ecologically sound and cost-effective decisions on crop health.

IPM has been strongly linked to processes which build on farmers' ability to learn, experiment and take appropriate decisions. Leaving behind more traditional "technology transfer" extension methods, IPM projects have developed around the Farmer Field School approach. Implemented first in Asia and then all over the world, Farmer Field Schools have proven to be a very effective tool for encouraging farmers to look for solutions to their problems, gaining the knowledge and the practical experience necessary to manage their farms successfully. IPM has therefore been built around farmers' own learning processes.

The Situation Today

On a daily basis, small scale farmers in different parts of the world still experience many of the difficulties reported twenty years ago. Although reliable statistics are difficult to find, the impact of pests and diseases in worldwide agriculture is considerable, lowering yields and overall production, resulting in losses that are equivalent to millions of dollars. These losses are not only unpredictable; they are also greater in fragile ecosystems.

Furthermore, the most common and widely available "solution" of using pesticides has only made problems worse. Farmer suicides as a result of pesticide debt traps have become common in countries as different as India and Ecuador. And pesticide poisoning is a common story, having reached endemic proportions in many countries. Farmers and labourers who are regularly in contact with pesticides suffer from severe health problems, seriously affecting them and their families. In Peru's village of Tauccamarca, as in many rural areas in Africa, Asia and Latin America, accidental intoxications have resulted in the death of children and innocent people. Indiscriminate use of pesticides has also resulted in the contamination of soils and groundwater, leading to the disappearance of fish and birds. Because of the lack of regulations, or of the difficulties in effectively enforcing them, these problems are greater in the rural areas of the nonindustrialised world.

Ecological Pest Management

IPM has had many successes, but the magnitude of the problem which farmers currently face forces us to look once again at pests and diseases and at their management. While it may be necessary to critically assess the extension methods followed during these last two decades, it seems equally relevant to stress and emphasise again one of the basic ideas behind the "different pest management system" which we called for more than twenty years ago: that pests and diseases are not an isolated part of agriculture, but rather a symptom of a broader problem, and need thus to be seen –and managed– accordingly.

Not surprisingly, this is known and recognised by many farmers. A few years ago in Tamil Nadu (India), at a meeting where farmers, NGO officials and government extension workers got together to discuss the most pressing problems in small scale farming, pests and diseases came out as the most important problem to be addressed. Farmers mentioned that the pesticides they used were increasingly ineffective, so that therefore they had to use more and more of them.

Expressing the views of most participants, one farmer stood up to say that "... we can try to tackle these pest problems, but we must understand that a plant suffering from pest problems is like a diabetes patient suffering from skin boils. The boils are a symptom of a deeper problem in the human body, and so are pests in the paddy crop a symptom of a deeper level health problem. We need to tackle the symptom first and then go deeper to address the cause of the problem. We have to look at the paddy plants and the soil and at everything around".

His views reflect the need for an ecological perspective when addressing the presence of pests and diseases in agriculture. Building on the wide range of experiences gained with IPM during the last two decades, an ecological pest management approach is one which focuses on managing pests as part of a wider ecosystem. As such, EPM is based on:

Minimising the disturbances which are caused by agriculture In whichever setting it is practised, agriculture will always cause disturbances and change the ecosystem, and one species (the crop) will nearly always dominate other plant species. These disturbances, however, need to be as small as possible. As Lanting (p. 6) points out, a successful pest management strategy is based on mimicking nature, redesigning a farm so that it resembles a complex ecosystem.

This will mean maximising a farm's positive ecological processes (such as nitrogen fixation, nutrient mineralisation); while at the same time minimising undesirable processes such as nutrient loss or erosion. In many cases it may be convenient to reduce tillage and thus achieve minimal soil disturbance; in many others it will be necessary to include perennial species and enhance a farm's overall diversity. The interaction of different species, as Amudavi *et al.* (p. 8) show, can have interesting results, contributing to the system's overall resilience.

Decreasing Plants' Vulnerability

Not all plants are equally susceptible to pests and diseases. This is even true for plants of the same species and variety: field observations show that pests prefer to attack plants under stress. Current studies, such as those falling under the theory known as trophosiosis, talk of a plant's "internal balance", directly related to its nutritional state. The best way to prevent the attack of pests and diseases is thus by providing a healthy and balanced environment and food supply. As Guazzelli *et al.* show (p. 14), there are many factors which can affect a plant's internal balance and thus lessen or increase its susceptibility to pest and disease attacks. These are related to the plant itself (such as adaptation to the local climate or its age) or to the environment (climate, light, temperature, humidity, wind). Plant vulnerability is also related to the different management practices which regularly take place in a farm, such as spacing, tilling, pruning, or the time of planting.

Understanding Pests and Acting Accordingly

The interactions between the components of an ecosystem vary greatly, and are specific to every location. In pest and disease management, one of the main considerations is the way the pest species behave: some show abilities to reproduce often and disperse widely, others are able to withstand competition or adverse conditions. To "know your enemy" (Belmain, p. 18) is thus a key strategy in every pest management approach. This knowledge needs then to be translated into action, considering, for example, the release of beneficial insects on a particular moment, adding bird nesting sites to a farm, or changing the sowing time of certain crops.

These principles are clearly visible in many traditional low input agricultural systems, where ecological principles form the basis for all pest management strategies. In brief, these refer to working with nature, and not against it. In agricultural terms, this means growing plants in the right soil, at the right time; nourishing the soil and relying on a system's biodiversity as a natural means to safeguard the whole system's health. Traditional wisdom is being maintained by many societies, while it is also being recreated in many "modern" farms. As shown by Reinders (p. 32), farmers in many of the intensive agricultural areas in the Netherlands have a very similar approach towards pests and diseases: this is not a separate problem which needs to be solved in isolation. Pests and diseases are dealt with by managing a farm as a whole. Managing a farm, however, and relying on its ecological processes, requires a thorough understanding of how these work.

Therefore EPM is, above all, based on farmers' skills, abilities and knowledge.

Building Knowledge

If EPM is based on farmers' understanding of their ecosystem and of the processes taking place in it, then training, education and knowledge building processes are essential. Many different participatory approaches for promoting sustainable agriculture have been developed, most of which work towards improving farmer decision making capacities and stimulating local innovations. Experiences showing the positive results of Farmer Field Schools, as a "model approach for farmer education", have been widely reported in the LEISA magazines.

Through Farmer Field Schools, farmers are trained to make an analysis of their agro ecosystem. In this way they become aware of the pest–predator balance and of the damaging effect of pesticides on such balance. They learn that it is better and more profitable to work with nature rather than against it. FFSs have become a very popular approach, taken up by NGOs and governments, on a small and a large scale. Their comparative advantage relies on a skilful incorporation of several principles: learner-centred, field-based, experiential learning; observation, analysis, assessment, and experimentation over a time period sufficient to understand the dynamics of key agro-ecological and socio-ecological relationships; peer-reviewed individual and joint decision-making based on learning outcomes; capacity building in leadership, social capital and empowerment.

The successes of the FFSs and other similar approaches show the truth of a common phrase: that knowledge is power. Understanding the ecological processes taking place in their farm not only helps farmers support and enhance such processes. It also helps to reduce the high degree of dependency many farmers have on chemical inputs and on the system –and vested interests– behind them.

Challenges

Ecological pest management is about bringing the balance back to disturbed ecosystems; it is also about learning to observe such balances. The enormous impact which pests and diseases have in today's agriculture forces us to pay special attention to these issues. Thinking of a cost-benefit analysis, for example, or on the need to easily visualise the impact of any pest management strategy on our health and on the environment, we need to continue developing tools to facilitate this learning process, building on the many successes seen.

However, as Schut and Sherwood show, the widespread dissemination of Farmer Field Schools during the last ten years needs to be reviewed with a critical eye. In many cases they need to be modified so that they really reflect farmers' needs and are based on their knowledge and interests. In many others, it has to be recognised that the approach has "eroded" as a result of complex social matters and opposing interests. It is becoming clear that until and unless these are addressed, Farmer Field Schools will not lead farmers to a lasting independence from the pesticide trap. They may easily fall back into old practices or, even worse, they may resort to GM crops and find out later that their dependence on pesticides has only increased.

Our main challenge is therefore to keep the collective learning spirit that has been built by approaches like the FFS alive. In all situations, farmers need to decide on how to control the incidence of pests and diseases in their specific context, relying on the natural balances within their ecosystem. There are no standard recipes or solutions available. Yet, there is much wisdom to be tapped, both "old" and "new". Old practices are being "validated" by modern scientists. The challenge is to apply and adjust this wisdom in each specific situation.

Bibliography

Arguello D. S.: *Production, Post-harvest Management and Exportation of Tropical Fruits in Costa Rica*, Wageningen, Technical Centre for Agricultural and Rural Cooperation, 1992.

Arora, Chandralekha : *Advances in Horticulture Science Research*, Oxford Book Company, Delhi, 2009.

Barbeau G.: *Tropical Fruits in Nicaragua*, Managua, Nicaragua Ministerio de Desarrollo Agropecuario, Agraria, 1990.

Bhattacharjee, Supriya Kumar : *Amenity Horticulture, Biotechnology and Post-Harvest Technology*, Pointer, Delhi, 2006.

Chadha K. L. and Pareek O. P.: *Advances in Horticulture: Fruit Crops,* New Delhi, Malhotra Publishing House, 1993.

Crucefix D.: *Avocado Variety Selection for Export Development,* Roseau, CARDI, 1996.

Currah L. and Proctor F. J.: *Onions in Tropical Regions*, Kent, Natural Resources Institute, 1990.

Dhatt A. S. and Singh Z.: *Propagation and Rootstocks of Citrus*, New Delhi, Malhotra Publishers, 1993.

Doijode S. D.: *Seed Germination in Fruits*, New Delhi, Malhotra Publishers, 1993.

Featherly H. I.: *Taxonomic Terminology of the Higher Plants*, USA, Iowa State College Press, 1954.

Gibbs, D.R. : *Chemotaxonomy of Flowering Plants*, McGill-Queen's University Press, Montreal, 1974.

Glowinski Louis: *The Complete Book of Fruit Growing in Australia*, Australia, Lothian Publishing Company Pty. Ltd., 1991.

Gunjate R. T. and Tawde A. B.: *Propagation and Rootstocks of Mango*, New Delhi, Malhotra, 1993.

Harlan, Jack R.: *The Living Fields: Our Agricultural Heritage*, Cambridge: Cambridge University Press, 1998.

Herklots G. A. C.: *Vegetables in South East Asia*, London, George Allen & Unwin Ltd., 1972.

Jeffers P.: *Evaluation of Four Onion Varieties in Montserrat*, Plymouth, CARDI, 1992.

Jerram, M.R.K. : *A Text-Book on Forest Management : In Agriculture, Horticulture and Forestry*, Asiatic Pub, Delhi, 2006.

Jugale, V.B. : *Horticulture Economy of Maharashtra*, Shruti, Delhi, 2004.

Malins A.: *Postharvest Handling of Pineapple and Mango*, Port of Spain, Trinidad and Tobago, 1992.

Matsuoka H.: *Cultivation of Panicum Genetic Resources for Evaluation of Characteristics*, Tokyo, JICA, 1997.

Mitra S.: *Postharvest Physiology and Storage of Tropical and Subtropical Fruits*, Oxon, CABI, 1997.

Morton Julia F.: *Fruits of Warm Climates*, Miami, Julia F. Morton Publisher, 1987.

Oldham P.: *Cost of Production of Major Tree Crops in Dominica*, Roseau, Ministry of Agriculture, 1991.

Peter, K.V. : *Basics of Horticulture*, New India, Delhi, 2009.

Pilgrim R.: *Post Harvest Handling of Minor Exotics*, St. George's, Grenada, 1996.

Prasad, Surendra : *Agro's Dictionary of Horticulture*, Agrobios, Delhi, 2003.

Sharma, V.K. : *Advances in Horticulture : Strategies, Production, Plant Protection and Value Addition*, Deep and Deep, Delhi, 2004.

Singh H. P. and Chadha K. L.: *Genetic Resources of Citrus*, New Delhi, Malhotra Publishing House, 1993.

Singh, D.K. : *Elements of Horticulture*, Agrotech, Delhi, 2005.

Singh, H.P. : *Challenges of Climate Change - Indian Horticulture*, Westville Publishing House, Delhi, 2010.

Singh, S.P. : *Advances in Horticulture and Forestry*, Scientific, Delhi, 2001.

Thavarasook C.: *Proceedings of the Mungbean Meeting 90, Chiang Mai, Thailand,* Bangkok, Tropical Agricultural Research Centre, 1991.

Weiss, R.F. : *Herbal Medicine*, Beaconsfield Arcanum, Beaconsfield, 1991 .

Whitwell A.: *Dominica Orchard Crop Management and Research Project: Pest and Disease Management,* United Kingdom, Natural Resources Institute, 1991.

Index

H

I

K

L

M

N

O

P

R

S

T

V

W

❑❑❑